FEMINISM IN THE WILD

FEMINISM IN THE WILD

How Human Biases Shape Our Understanding of Animal Behavior

AMBIKA KAMATH AND MELINA PACKER

The MIT Press
Cambridge, Massachusetts
London, England

The MIT Press would like to thank the anonymous peer reviewers who provided comments on drafts of this book. The generous work of academic experts is essential for establishing the authority and quality of our publications. We acknowledge with gratitude the contributions of these otherwise uncredited readers.

This book was set in Adobe Garamond Pro by New Best-set Typesetters Ltd. Printed and bound in the United States of America.

Library of Congress Cataloging-in-Publication Data

Names: Kamath, Ambika, author. | Packer, Melina, author.
Title: Feminism in the wild : how human biases shape our understanding of animal behavior / Ambika Kamath and Melina Packer.
Description: Cambridge, Massachusetts : The MIT Press, [2025] | Includes bibliographical references and index.
Identifiers: LCCN 2024023049 (print) | LCCN 2024023050 (ebook) | ISBN 9780262049634 (paperback) | ISBN 9780262382281 (epub) | ISBN 9780262382298 (pdf)
Subjects: LCSH: Animal behavior.
Classification: LCC QL751 .K26 2025 (print) | LCC QL751 (ebook) | DDC 591.5—dc23/eng/20240911
LC record available at https://lccn.loc.gov/2024023049
LC ebook record available at https://lccn.loc.gov/2024023050

10 9 8 7 6 5 4 3 2 1

To Richard Levins and Richard Lewontin, for writing the book that made Ambika question everything, in the best way.

To Pepper, for being the dog who made Melina question everything, in the best way.

It is remarkable how Darwin recognizes among beasts and plants his English society.
—Karl Marx, in an 1862 letter to Friedrich Engels

. . . what a celebration when we realize that our survival need not make us into monsters.
—Alexis Pauline Gumbs, *Undrowned*

Contents

Figure 1.1
Coho salmon releasing their eggs and sperm in the Quinsam River in British Columbia, Canada. Original in color. Photo: Eiko Jones.

1 DISMANTLE BY BUILDING DIFFERENTLY

SALMON, TWO WAYS

Let's imagine ourselves standing by a stream somewhere in the Pacific Northwest of North America, bending our heads over the water to watch some Coho salmon. We don't yet know much about these fish, but we can start to describe what we observe: the fish's morphology (that is, what they look like and how their bodies are structured), and their behaviors. One of them is a bit bigger than the others, with a wide strip of bright red running down the length of their flank and a large, curved snout, colored jet black except for a distinct white stripe. This fish swims toward another fish, whose back is a silvery brown speckled with black, and whose sides are blotched with fainter patches of red. The second fish sometimes flashes yellow while flipping on their side, and thrashes their tail vigorously along the riverbed—the fish appears to be digging. In time, other fish swim nearer to these ones—some larger and more brightly colored, others smaller and duller. There's more swimming closer to and farther from one another, there's more digging, and eventually the fish release clouds of milky liquid into the water, their mouths agape as they do so. With so many fish in view, it's hard for us observers to know precisely which of the fish have just released their eggs and sperm into the water. We're watching Coho salmon group sex, and it seems to be a pretty complicated affair (figure 1.1).

The one fish who we saw digging in the riverbed is also the one who releases eggs—for now, let's call her a female. And the fish who release sperm (males, for now) can be divided into two types: the bigger, redder fish with the larger, black, hooked noses are known, straightforwardly, as "hooknoses," while the smaller fish who are silvery brown all over are known as "jacks." The groups of salmon populating this stream vary in composition. Some groups are made up of a female and just one hooknose, but many have a mix of jacks and hooknoses; every now and then, we see a female grouped with only a jack or two, no hooknoses.

To categorize Coho salmon into females and two types of males, and to describe their interactions, is to begin answering the questions "Who are these fish?" and "What are they doing?" However, the answers to these questions do not yet constitute what many people would think of as a complete scientific understanding of Coho salmon behavior, because they do not include an explanation for *why* these fish behave in the ways they do.

Scientific answers to *why* questions about animal behavior are rooted in the theory of adaptation by natural selection, which was first formulated by Charles Darwin in 1859.[1] The theory of natural selection describes a *process*—natural selection—which Darwin proposed to explain a *pattern*—adaptation. Anywhere we look in nature, we seem to find uncanny alignments between organisms' biology and the conditions of their lives. We might notice the near-perfect match in length and curvature between the bill of a hummingbird and the flower whose nectar he feeds on, or the precise match in color and pattern between a peppered moth's wings and the bark of the trees she rests upon. Biologists call these sorts of close matches *adaptations*, and adaptations seem to demand explanation. The theory of adaptation by natural selection states that for all organisms, those individuals whose biology best fits their environmental and social conditions are most likely to survive, reproduce, and pass on their biological characteristics to the next generation. Over many generations, populations of organisms evolve to *adapt* to their conditions, leading to the evolution of traits that are adaptations.

Thus, to understand the evolution of an animal's behavior, biologists pay careful attention to how that behavior affects the animal's survival and reproduction. This combination of survival and reproduction is referred to as *fitness*. To understand *why* Coho salmon behave as they do during group sex, then, one must try to connect their mating interactions—the release of eggs and sperm into the water while in proximity, and the events leading up to it—to their fitness.

In a scientific paper published in 2005, an animal behavior scientist named Jason Watters proposed one such connection.[2] In Watters's telling, female Coho salmon prefer their eggs to be fertilized by sperm from jacks, and behave in ways that make such fertilizations more likely. Females' mating preferences thus bring about the jacks' reproductive success, which is a key component of fitness. However, in seeking out the jacks, females end up being unable to avoid the large, aggressive hooknoses. Being smaller, the jacks cannot do much to fight against the hooknoses, but they can do their best to release their sperm when the females release their eggs, even if there also happens to be a hooknose around, releasing his sperm too. In this story, the jacks are described as *cooperators* and the hooknoses are described as *coercers*.

It turns out that Watters's description of Coho salmon mating behavior is unusual. In most of the stories that get told about how male animals compete for opportunities to mate with females, the little guys are not the females' preferred mates—the big guys are. In the mainstream and more familiar story about Coho salmon too, the big guys win, mostly. The hooknoses are typically described as *fighters*, who compete hard against other hooknoses, biting at their rivals and chasing them away from the female as she digs her nest and prepares to release her eggs.[3] The jacks, by contrast, are more often described as *sneakers* who weasel their way into mating interactions between females and hooknoses. One can see how the jacks' less vivid coloration, and how they hang back in the shadows while the hooknoses fight among themselves, might reinforce the impression that the jacks are sneaky.

In both these stories of Coho salmon mating, both jacks and hooknoses end up with some reproductive success, and some fitness, which means that both stories are evolutionarily plausible. The stark difference between the two stories revolves around female agency and choice—do the females *want* to mate with the jacks or with the hooknoses? The answer to this question places the two types of males in different light: if the females prefer jacks, the hooknoses are *coercive*, whereas if the females prefer hooknoses, the jacks are *sneaky*. Of these two stories, the latter is far more widely accepted, and continues to be the story told by most animal behavior scientists studying Coho salmon, Watters's 2005 study notwithstanding. For example, a 2023 paper by fisheries scientist Erika King and colleagues described these fish's mating dynamics as such: "[Hooknose] males typically maintain access to females by fighting and guarding. Jacks, on the other hand, . . . achieve spawning success by taking up satellite positions around the female then sneaking in to release their sperm as the eggs are being laid instead of fighting."[4]

Which stories scientists tell about Coho salmon mating matters to how humans monitor and manage these fish's populations. Under the mainstream story, it's easy to overlook the jacks. "The contribution of jacks is often ignored: their abundance is not always monitored and when it is, they are prone to being undercounted," King and colleagues wrote.[5] In hatcheries, jacks are sometimes excluded from mating altogether. The jacks' contributions to the reproductive dynamics of Coho salmon populations are readily disregarded when they're thought of as sneaky. But it would be much harder to overlook them if jacks were seen as the females' preferred mates.

You would think that data on these fish's actual behavior ought to rule in favor of one story over the other, and, at first, that seems possible. For example, Watters documented that females dig more (in preparation for laying their eggs) when they're interacting with jacks than with hooknoses, and spend more time releasing their eggs when there's a jack in the group of males releasing sperm, compared with when the group is made up of only hooknoses. One could interpret that as females investing more into reproduction when jacks are around, meaning that females prefer jacks. (Watters's conclusions rely on some assumptions, namely that more time spent digging actually corresponds to a better nest, and that more time spent releasing eggs also means that more or better eggs are released.)

But equally, one can interpret these data as being consistent with the more mainstream story, that females prefer hooknoses. Maybe females dig more when jacks are around because they're biding their time, waiting for a hooknose to show up. And maybe they release more eggs when a jack is nearby as something of a throwaway: yes, the female cannot prevent a jack's sperm from fertilizing some of her eggs, but by releasing more eggs, she can do her best to ensure that a larger number of her eggs will nonetheless be fertilized by hooknose sperm.

Other data from Watters's study are harder to explain from within the mainstream story of hooknoses as fighters and jacks as sneakers. For example, he found that females often start releasing their eggs *after* a hooknose has tried unsuccessfully to chase a jack away, when both the hooknose and the jack return, but not after unsuccessful chases of other hooknoses. Yet other data are equally consistent with both stories; females spend a majority of their time in proximity to hooknoses, not jacks, which could equally mean that hooknoses are the females' preferred mates or that hooknoses are coercive harassers whom the females cannot escape. So while one might imagine that data would rule definitively in favor of one hypothesis or another—that is how science is supposed to work, after all—in fact it's easy to interpret many different details to support either story.

At this point, if you know anything about salmon, you might find yourself saying, "But we *know* that the mainstream story is true! We've known this for years, there's so much research supporting it, and the extent of Watters's evidence is tiny in comparison!" Consider, however, that if the mainstream story is the *only* story that scientists have seriously considered so far, even the existence of vast amounts of research on that story doesn't necessarily mean that it's true. Prior to Watters's 2005 study, the possibility that Coho salmon jacks might not be sneakers was, in all likelihood, never seriously considered. Even after 2005, researchers have not directly addressed the question of which of these two stories is "more" true, and the mainstream story has persisted as the accepted explanation for how Coho salmon mate.[6] We can't yet say whether jacks are sneaky or not, because most research on Coho salmon simply *assumes* they are sneaky. When our certainty about the prevailing story depends on never having seriously considered other possibilities, how certain can we actually be?

TO BE OR NOT TO BE . . . TERRITORIAL

Anoles are bright-eyed, tree-dwelling, insect-eating lizards found in the southeastern United States, Mexico, and throughout much of Central and South America. Walking through forests in these regions, it's often easy to spot anoles by eavesdropping on their conversations with one another.

Figure 1.2
A green anole displaying an extended dewlap on a palm tree in Mosquito Lagoon, Florida. Photo: Ambika Kamath.

They use body movements—pushups, head bobs, and the rapid extension of a brightly colored, fan-like flap under their throats called a "dewlap"—to communicate (figure 1.2). For over a hundred years, biologists have made sense of these conversations, and of the social and reproductive lives of anoles more generally, using the framework of territoriality.

According to the framework of territoriality, animals occupy areas called territories and defend them against intrusions by their competitors. Furthermore, males' territories are often expected to contain the territories of one or more females; by defending their territories against intrusions from other males, the males who are so-called owners of territories can hope to be the sole mate of the females contained within. When Ambika began studying anole behavior, she took the territoriality framework for granted. But in time, she noticed that the lizards were behaving in a myriad of ways that this framework didn't capture. Ambika was far from the first to notice these anomalies—previous research had documented how both male and female individuals frequently moved far from the areas they were expected to stick to, and that female anoles very often mated with more than one male. Nonetheless, the territoriality framework continued to hold sway.

"The framework simply isn't realistic!," Ambika found herself complaining, to anyone who would listen. A friend of hers, once an avid biology student and now a businesswoman, heard this complaint and asked, "Wait, why do you scientists think these lizards behave like they live in Victorian England?"[7] Why, in other words, were female lizards thought to be the passive property of male lizards rather than active agents of their own accord? This single comment changed the course of Ambika's research. It sparked the idea that the earliest scientists studying anole behavior—largely, if not exclusively, rich, white, heterosexual, male, and American in the early twentieth century—had perhaps latched onto the territoriality framework because it aligned with their social norms, including a preference for property ownership and a distaste for female promiscuity.[8]

The word "territory" is so familiar, both within and outside of biological contexts, that it seems not to need defining. Indeed, that lack of definition was a large part of the problem. While investigating the earliest studies of anole lizards, published in the 1930s, Ambika learned that scientists regularly described anole lizards as territorial without explaining what they meant by that. Moreover, many scientists based their assertions of territoriality in anoles on limited, even strange, data.[9] For example, the authors of the very first observational study of these lizards' social behavior remarked that, in their lab populations, close to half of the copulations they saw were between males.[10] Sex between male anole lizards had, at the time, not been recorded in the wild and so the authors concluded that, in nature, these male-male copulations must be prevented by *something*, and that *something* must be the lizards' tendency toward territoriality. Other early studies were equally certain that anoles are territorial. Some reached this conclusion based on observations of the locations of individual lizards over time, which are the *kind* of observations that contemporary scientists also use to investigate territoriality. But the *volume* of data collected by early scientists was about an order of magnitude smaller than what scientists today would consider acceptable for supporting any sort of conclusion. And some early scientists based their conclusions on even more questionable observations; one researcher concluded that a wild anole lizard was territorial simply because the researcher saw the lizard display his throat-fan at a jar filled with live lizards of a different species, which the researcher had happened to leave on the steps of his back porch.[11]

On the basis of these early and often flawed studies, the idea that anole lizards are territorial took hold, and most later research on these animals ended up designing studies that *assumed* territoriality rather than *testing* it. It thus became unlikely that research on anole lizards would even detect, let alone take seriously, social and reproductive behaviors that were not territorial. Put simply, when one assumes territoriality, one finds territoriality. Most scientists studying anole lizard territoriality never seriously considered the possibility that territoriality might not be a great framework for understanding these animals' social behavior. Unless they're explicitly challenged, and sometimes even after that, prevailing stories such as the sneakiness of Coho salmon jacks or the territoriality of anole lizards tend to stick.

WHAT SCIENCE IS AND CAN BE

The stories told by scientists shape scientific imaginations and realities. They determine the questions that scientists think are interesting, the research that they try to get funded, the experiments they design, the data they collect, how they interpret the data, and what conclusions they reach—conclusions which scientists then communicate in books, podcasts, and articles as truth and fact. And when much or all of the research on, say, Coho salmon mating dynamics or anole lizards' social interactions has been conducted from within the intellectual, imagined world created by a single narrative, then it becomes very difficult to know whether that particular story is actually an illuminating way of describing that behavior. In other words: when we tell just one story, we don't know what we don't know.

The point here is *not* to adjudicate which specific story about salmon mating (jacks as sneakers or jacks as cooperators) or anoles (territorial or not) is the "most correct." The point is not even about Coho salmon or anoles or any specific animal. The point is that, within the realm of animal behavior science, it is *possible* to tell different stories about what animals are doing and why. Allowing for a multitude of stories rapidly expands the horizons of our scientific imaginations, by encouraging us to consider what the prevailing stories might be taking for granted.

To continue with the fish example: although the two stories about Coho salmon are starkly different from one another in some ways, in other ways they are quite similar. For one, they are both rooted in the assumption that males compete for mates and that females have a single preference for one of the two types of males, with the other type of male somehow acting in opposition to the females' desires. Can one imagine a third or fourth or fifth description of Coho salmon mating, in which conditions of competition and uniformity are *not* assumed? In doing so, scientists might move away from wondering about the relative reproductive success of jacks and hooknoses or asking how they compete, and instead wonder whether there are lots of different ways to be a salmon who is mostly okay at surviving and reproducing. Scientists might search for stories in which different females prefer different types of males, or simply have no discernible preference at all. Scientists might accept that females are perfectly okay mating with multiple males of either type, with no explanations of coercion or sneaking required. They might stop thinking of jacks and hooknoses as two types of males, and instead start thinking of Coho salmon as having three sexes. To get comfortable with this multitude of possibilities is to let go of the idea that there is a single, "true" scientific story about any given animal. *Feminism in the Wild* is a book about what happens to the science of animal behavior when we embrace this multitude of possibilities.*

---

* A quick note about the word "animal." Of course, humans are animals too, and the very category of "animal" as distinct from humans presents a problem because it suggests that a huge diversity of organisms, excepting humans, can be lumped into one, undifferentiated group. Language that reinforces the human/animal binary by positioning humans as separate from other animals also positions humans as *superior* to other animals. However, the alternative terms "more-than-human animal" or "nonhuman animal," and "human animal" are undeniably clunky. For ease of reading then, we use the terms "human" and "animal" throughout this book, with more than a little ambivalence.

It may seem antithetical to the notion of science to make room for multiple interpretive stories about the same observations of nature without then trying to ascertain which story is right and which is wrong. After all, isn't the point of science to get us ever closer to understanding what is objectively true about the world? Can multiple scientific truths about ostensibly the same thing really coexist? It's worth pausing here to consider what people typically mean by "science," referred to as "mainstream science" throughout this book. Usually, "science" refers to European philosophies and methods for understanding the world that emerged during that region's "Age of Enlightenment" (1685–1815). The scientific method that emerged in this particular place and era was, and still is, widely considered to be uniquely objective and neutral, and thus believed to be a superior way of ascertaining the truth, as compared to other cultures' knowledge traditions.

The crux of the argument in *Feminism in the Wild* is that mainstream science is not as objective, neutral, or singularly truthful as people might expect. A potentially controversial claim, to be sure, but one that is supported by decades of research in feminist science studies. The foundation of this claim is made up of two facts. One: all of our worldviews, including the scientific worldviews of scientists, are shaped by our social locations—who we are, where we come from, what we do, and the kinds of social, economic, and political power we have. Two: both historically and currently, practitioners of mainstream science tend to come from the more elite, powerful ranks of society. These two facts together imply that mainstream scientific narratives tend to align with the social, economic, and political narratives of those who hold power in our societies. Moreover, because mainstream science is thought to reveal objective truths about the natural world, the narratives of the powerful come to be seen as just how nature works. In other words, the perspectives of the powerful become *naturalized*, making it much harder to escape these uneven power relations or dismantle the narratives they produce. Meanwhile, the perspectives of anyone other than the elite remain at the margins of scientific understandings of nature.

Feminism in the Wild dismantles the foundations of mainstream animal behavior science by showing how these foundations are inextricable from, and naturalize, the systems of power that dominate our world today:

patriarchy, racism and colonialism, homophobia and transphobia, ableism, and capitalism. This book synthesizes decades of research from a plethora of scholars who are, in various ways and to different extents, critical of mainstream animal behavior science. Such critical research can be found not only in the field of feminist science studies but also within animal behavior science itself, as well as the closely related fields of behavioral ecology and evolutionary biology. Ambika is trained as a behavioral ecologist and evolutionary biologist, and Melina is trained as a feminist science studies scholar—this collaboration brings our two fields of expertise into conversation with one another.

Feminism in the Wild is, of course, informed by us authors' own political perspectives, which are anti-patriarchal, anti-racist, anti-colonial, anti-ableist, anti-capitalist, and queer. We think of these political commitments collectively as *feminist*. In adopting this expansive view of feminism, we follow the lead of activists and scholars who have interrogated the origins of seemingly immutable categories such as gender or race, and have insisted on an understanding of feminism that is deeply intersectional, or even multi-dimensional.[12] In other words, not only does a person have multiple social identities and occupy different social locations in different contexts, but also a person's identities and locations *cannot be understood independently of one another.*[13] This means, to us, that feminism is not simply about gender equity but rather about dismantling all forms of oppression.

Being explicit about how science is intertwined with political values, feminist or otherwise, does not invariably lead one to commit the naturalistic fallacy, which is the idea that humans are supposed to do (or cannot help doing) what other animals do, simply because such behaviors are "natural." As much as any of us may want to believe that animals live in capitalist hellscapes rife with scarcity and competition or in feminist utopias free of violence and inequality, we all have to accept that nature does not care about what we want. As historian Gregg Mitman put it: "We must stop looking to nature for reassurances about humanity, for we will inevitably see . . . a reflection of what we want to see."[14]

So we cannot pretend that science will solve our political disputes, and we cannot pretend that science isn't political. What, then, do we do?

Feminism in the Wild argues that all we can do is remain alive to the political complexity of science, and fully accept the fact that scientific ideas are always intertwined and interacting with social, cultural, and political beliefs. The point of accepting this intertwining is never to paint our moralities onto animals uncritically. Instead, the point is to become curious about the entanglements of scientific stories with the human contexts in which these stories are constructed, and to interrogate the material effects and political stakes of endorsing one scientific story over another. Our goal in writing this book is to show that the practice of science itself can make room for such curiosity. For example, the work of science could include examining changing descriptions of male Coho salmon reproductive strategies in light of evolving ideas of masculinity in the late twentieth-century United States, or teasing apart the ways in which notions of territoriality in anoles conform to ideas about human property ownership.[15] *Feminism in the Wild* contends that engaging in such expansive inquiries will bring us closer to understanding animals and closer to justice than current, mainstream approaches to animal behavior science ever could.

The beauty of a feminist approach to science is that it does not regard the fundamental interdependence of who scientists are and the science they produce as a problem. Rather, feminist approaches to science insist that the most effective way to understand the world is to regard it through many different lenses and to hold, all together, the many different views of the world that result. The feminist reimaginings of animal behavior science in the pages ahead are multifaceted, messy, and full of contradictions, which isn't quite compatible with what many people believe science is or needs to be; nevertheless, it is a more honest and accurate reflection of the messiness of life. Feminist science is only possible when scientists embrace the wild instead of trying to tame it into the prevailing narrative.

FREEING THE FROGS

For the last thirty years or so, well-intentioned toxicologists and environmental scientists have been expressing serious concerns about the harmful effects of chemical pollution on frogs.[16] At the start of his career, environmental scientist Max Lambert wanted to be part of this effort to better protect frog habitats from toxic chemicals. "It just so happened," he told us, "that the chemicals [my supervisor] was looking at . . . were related to sexual development."[17] Lambert did not set out to become a frog sex scientist, but toxicological methods led him there nonetheless.

In the late 1990s and early 2000s, toxicological research tended to report "sex-reversed" frogs (males who had become females) and "intersex" frogs (individuals with egg-producing cells in their testes, for example) in highly polluted environments, whether simulated in laboratory experiments or found in the wild. Researchers concluded that chemical exposures *cause* "sex reversal," leading to skewed sex ratios—too many females!—in wild frog populations. Other biologists also noted male-male sexual encounters among chemically exposed frogs (primarily in the laboratory), and interpreted this same-sex sexual behavior as evidence of chemically induced harm.[18] Lambert's 2010s research on American green frogs (figure 1.3) fit right into this pattern. He found that frog populations in suburban ponds

Figure 1.3
An American green frog. Photo: Max Lambert.

had more females than males compared to frog populations in more rural ponds. To a toxicologist, the obvious explanation for this pattern would have been chemically induced sex reversal in suburban ponds. Like others before him, Lambert could have assumed that suburban ponds were more polluted than rural ponds, concluded that chemical pollution was "feminizing" male frogs, and left it at that.

But Lambert didn't leave it at that. He noticed that, although it was true that there were more female frogs in suburban ponds, these ponds actually had *less* skewed sex ratios than rural ponds. The difference in sex ratios he'd observed stemmed from the fact that rural ponds had *male*-biased sex ratios—too many males! Lambert became curious about how these different sex ratios came to be, and started to track frog sex ratios at the egg, tadpole, and adult stages in suburban and rural ponds.

What Lambert found was entirely unexpected. Regardless of chemical exposure, American green frog tadpoles were "just switching sex pretty much everywhere," which could lead to the presence of egg-producing cells in testes.[19] Sex changes at the tadpole stage appeared to result from temperature changes rather than exposure to hormone-interfering chemicals. Frog populations' "surplus" of females, meanwhile, seemed to have more to do with higher rates of male mortality than male-to-female sex changes.[20] What is more, intersex frogs could still produce offspring, contrary to concerned biologists' claims that the presence of "transgender" frogs necessarily signaled species decline.[21] As Lambert put it, tadpoles that switch sex are "making it to adulthood, becoming sexually viable, and reproducing successfully. And that's interesting because a lot of research on . . . sexual abnormalities suggests those animals are inferior, they can't mate, no one wants them, they're going to be unloved for the rest of their lives. [But] they're probably actually doing just fine out there and you couldn't really tell the difference anyway."

Meanwhile, and unbeknownst to Lambert, a feminist science studies scholar—none other than Melina!—was engaged in a queer feminist analysis of the science of toxicology. Through her work, Melina became increasingly fascinated, and disturbed, by the unarticulated assumptions undergirding toxicological experimentation and anti-toxics advocacy alike, especially surrounding hormone-interfering chemicals and their purported effects on sexualized behaviors. For example, toxicologists assumed that male laboratory rats engage in more "aggressive play" behaviors than their female counterparts because, well, boys will be boys. Toxicologists next determined that a male rat exposed to an estrogenic toxicant had been "feminized" (which the toxicologists interpreted as abnormal and harmed) because he appeared to engage in less aggressive play post-exposure. The extrapolation followed, then, that male humans will be feminized by exposure to estrogenic chemicals, and that this feminization is a bad thing.[22]

Melina's queer feminist analysis of this kind of research raised many questions. Why were scientists assuming that sex and gender are binary and moreover that male = masculine = aggressive, while female = feminine = passive? What does "aggressive play" even mean, and how can one recognize it? How can one generalize from a captive, laboratory rodent's highly constrained behaviors (not to mention their highly contrived genetics) to wild rodents, never mind to humans? And what's so threatening about supposedly feminized (human) males anyway, especially if feminization means less aggression? Melina wasn't arguing that we should *not* be concerned about highly toxic substances permeating our environments and harming our health without our consent, all to chemical corporations' great profit. But she was certain that there had to be a way to conduct research and express alarm over toxic environmental pollution without naturalizing and normalizing binary sex categorization and stereotypical human gender roles across different animal species.

An eventual cross-disciplinary collaboration between Max (Lambert) and Melina—they're on a first-name basis now—made more explicit the queer approach that Max's previous research had, in some ways, implicitly adopted.[23] It was an implicitly queer feminist approach that led Max to question several assumptions that well-meaning biologists had previously left unquestioned, namely: that queer sex among frogs is unnatural, that tadpole sex changes are unnatural, and that intersexed frogs are not only unnatural but also cannot reproduce. In other ways, however, Max's earlier research on frog sex was squarely situated within dominant heterosexist narratives. For example, he used terms such as "demasculinized" frogs, which pathologize variation in sexual expression, and are unfortunately standard in his field. But ultimately, Max remained open to alternative possibilities for frog physiology and sexual behavior.

A queer feminist approach to frog sex would not immediately assume that changes in frog sex are bad things in and of themselves, and would instead work to uncover how chemically polluted waters *are* harming organisms (including the people, animals, and plants who ingest the water through osmosis, drinking, eating, swimming, and more).[24] A queer feminist approach also would not frame chemically poisoned frogs as pitiable freaks of nature, as well-intentioned scientific and media coverage tends to do.[25] Importantly, a queer feminist approach does not dismiss the reality that chemical exposures are indeed harmful to frogs, and may even harm reproductive health. Indeed, by paying close attention to this very reality, scientists can focus on the effects of toxic chemical exposures that are demonstrably life-threatening (such as DDT's weakening of raptor egg shells) while linking those chemical exposures to specific chemical manufacturers' profit margins and specific governments' lax regulations. Scientists taking this queer feminist approach also have reason to renew their attention to animals' basic biology, asking questions like: How and why does temperature influence tadpoles' sexual expression? Why are male frog mortality rates higher than those of females? And, given that sex changes do not necessarily preclude reproduction but might alter it in as-yet-undocumented ways, what are the implications of global warming for frogs thriving well into the future?

Max's openness to the messiness of life took him from being someone who uncritically aligned himself with the mainstream scientific story that male frogs are "feminized" by hormone-interfering chemicals to someone who embraces queer feminist possibilities for frog physiology and sexual behavior, and subsequently finds data that support alternative explanations, thus adding to our knowledge of basic biology. It's this openness to possibility about the natural world that we authors hope you will carry with you as you read this book, and long after. As Max put it: "Be open to surprises . . . biology will surprise you. Let the animals tell you something different about the world than what you thought."[26]

Figure 2.1
A fruit fly. Photo: Ben de Bivort.

2 WHO WE ARE AND WHAT WE KNOW

THE MORASS OF FEMALES MATING WITH MULTIPLE MALES

In 1948, the botanist Angus J. Bateman made an unusual foray into the world of animal sex, publishing a paper that reported the results of a mating experiment on fruit flies (figure 2.1).[1] In this experiment, Bateman arranged between six and ten fruit flies, males and females, into little vials, where they mated with each other and produced offspring. Fruit flies in the experiment had distinct genetic mutations that led them to look different from one another: some had hairy wings, others had no eyes, and yet others had shortened bristles, among other variants. Because these mutations were inherited, offspring often looked like their parents, which helped Bateman identify the parents of many of the fruit fly babies.[2] With these data, he could then estimate the relative reproductive success of each adult fruit fly, asking which flies bore more offspring, and which fewer. He could also estimate the number of males that each female had mated with, and vice versa.

Bateman's inspiration for carrying out this experiment had been Charles Darwin's pronouncements, almost eighty years prior, about the dynamics of males and females during reproduction. In 1871, Darwin proposed a theory of sexual selection.[3] Whereas *natural* selection was mostly concerned with the traits that improved survival, *sexual* selection focused on traits that improved reproductive success. Darwin had argued that sexual selection would lead males to be "eager" to mate, each attempting to mate with as many females as possible and thus sire as many offspring as possible (males were imagined to have effectively infinite supplies of sperm). Bateman's experiments generated the data needed to assess this association between how many mating partners an individual has and how many offspring that individual produces. Based on Darwin's theory, Bateman predicted that males who mated with more females would also sire more offspring.[4]

Darwin had different expectations for females than for males; he thought that females would be as selective as possible, choosing among the males and mating only with the best. According to Darwin, females' relatively limited egg supply necessitated this strategy of quality over quantity. He therefore posited that sexual selection would lead females to be "coy," and so Bateman predicted that females who mated with multiple males (or *polyandrous* females) would *not* have more offspring than females who mated with just one male (*monandrous* females). Or, put in evolutionary terms: there would be no selection for females to mate with multiple males.[5]

To test his predictions, Bateman plotted his data on a graph, putting the number of mates on the horizontal axis and the number of offspring on the vertical axis, with separate lines for males and females. He expected the line for males to rise upward—more mates, more babies. In contrast, he expected the line for females to plateau or even dip as their number of mates increased above one, giving females no added evolutionary advantage to mating with more than one male. Inexplicably, instead of plotting all his data on a single graph, Bateman split his results into two separate graphs, noting only that the first four iterations of his experiment "differed somewhat" from the last two.[6] (This sort of after-the-fact decision is, by the way, statistically unjustifiable.)

One of Bateman's two graphs, the second one, went on to become the most important piece of evidence in support of projecting Darwin's Victorian-era ideals of masculinity (eager, profligate males) and femininity (coy, choosy females) onto animals in general. The second graph conformed to Bateman's predictions, based on Darwin's theory: males that mated with more females had higher reproductive output, but female reproductive output did not increase when they mated with more males. Ignoring the first graph and focusing on the second, Bateman concluded that promiscuity pays off for males but not for females. He saw simply no reason for females to mate with more than one male, and had no problem extending this result from some of his fruit flies to all animals, writing, for instance, that "the greater dependence of males for their fertility on frequency of insemination . . . is in fact an almost universal attribute of sexual reproduction," and that "there is nearly always a combination of an undiscriminating eagerness in the males and a discriminating passivity in the females."[7]

Bateman's paper languished in near-obscurity until 1972, when evolutionary biologist Robert Trivers brought renewed attention to that second graph, building upon it a whole theory of sexual dynamics that only further reinforced the notion that male and female animals conform to Victorian-era gender stereotypes wherein "the males are almost always the wooers."[8] As part of his argument, Trivers reported that Bateman had observed males vigorously courting females and females rejecting said courtship, even though Bateman's experiment included no observations of the flies' behavior at all.[9] Trivers's argument caught on quickly, however, inspiring a huge body of scientific research on sexual behavior in animals and humans that took patriarchal stereotypes as a given. As writer Angela Saini put it in her book *Inferior*, an investigation of gender bias throughout science: "Bateman's theories, once almost forgotten, were transformed into a fully blown set of universal principles, cited hundreds of times and considered as solid as a rock. On that rock now rests an entire field of work on sex differences."[10]

Recall that this enormous explosion of scientific support for patriarchal gender stereotypes grew from Bateman's *second* graph alone. What about the first graph? The first graph showed a different result, namely that female reproductive success *also* increased when they mated with more males—the opposite of what Darwin would have predicted. There was no scientific reason to ignore the first graph, and if Bateman, or Trivers, had paid appropriate attention to this first graph, it's possible that biologists who subsequently built upon their work might not have expected females to be coy and choosy, and instead would have expected females to mate with multiple males. Why? Because if mating with multiple males is associated with higher reproductive output in females, then *polyandry* (mating with multiple males) is simply a result of sexual selection favoring those behaviors in females that increase their fitness. (Recall that "fitness" refers to the combination of survival and reproduction; Bateman measured the fruit flies' reproductive output as a proxy for fitness.) In other words, if scientists had taken the first graph seriously (or combined the data from the two graphs), they would have expected polyandry to be the norm, and *monandry* (females mating with just one male) to be the anomaly requiring explanation, not the other way around.[11] Remarkably, one can actually reach the same conclusion from the second graph too. Yes, the second graph showed that a female's fitness doesn't increase when she mates with more than one male, but *it doesn't decrease either*, so scientists could have just as readily concluded from the second graph that a female's fitness is unaffected by how many mates she ends up having.[12]

For science to have considered polyandry to be unsurprising, Bateman and Trivers would have had to swim against the tide of the patriarchy. Perhaps unsurprisingly, these men did not do so. When asked in 2001 why he focused on the second and not the first of Bateman's graphs, Trivers reportedly said "unashamedly that it was pure bias."[13] But this is not to say that *no one* swam against the tide of the patriarchy—in fact, quite the opposite. Not long after Trivers's publication, scientists and science studies scholars, most of whom would identify as feminist, strongly rejected what came to be known as Bateman's Principles. As feminist evolutionary biologist Patty Gowaty described, these scholars' "feminist consciousness nurtured the recognition of alternative hypotheses and underlying assumptions of reigning theories."[14] Their analyses not only called out the "close conformity between [Bateman's Principles] and post-Victorian popular prejudice" but also reported data from myriad different animal species that do not follow Bateman's limited edicts for female behavior: lionesses who mate a hundred times a day while sexually receptive, female baboons and chimpanzees who initiate courtship with males, and female fish who seek out copulations with multiple males long before they are fertile, storing sperm for eventual fertilization, for example.[15] As feminist animal behavior scientist Zuleyma Tang-Martínez summarized in 2005, "the last thirty-five years have seen an abundance of studies confirming that females in many species . . . are sexually assertive, actively seek multiple copulations, and routinely mate with more than one male."[16] In other words, not what Darwin, Bateman, or Trivers concluded at all. The fact that mainstream animal behavior scientists could ignore such examples underscores how easily science is shaped by scientists' social and political locations. As primatologist Sarah Blaffer Hrdy, a leader of the charge against Bateman's Principles, put it: "When generalizations persist for decades after evidence invalidating them is also known, can there be much doubt that some bias was involved? We were predisposed to imagine males as ardent, females as coy. . . . How else could the [fruit fly] to primate extrapolation have entered modern evolutionary thinking unchallenged?"[17]

After decades of research pushing back against Bateman's Principles, in the early 2000s, Gowaty went back to the source of it all. Along with colleagues, she analyzed and attempted to replicate Bateman's original experiment. Even beyond Bateman's baffling choice to split the data into two separate graphs and focus on only one of them, Gowaty and colleagues uncovered an astounding array of confusing choices, understandable shortcomings, and outright mistakes in Bateman's 1948 paper, rendering his results entirely unreliable.[18] You'd think this would be the final nail in the coffin; as science writer Lucy Cooke put it: "The fact that Bateman's [Principles are] not supported by Bateman's data would seem to be something of a terminal setback, empirically speaking."[19]

But despite decades of feminist intervention, the science of animal sex remains attached to its founding misconceptions, and Bateman's Principles remain at the center of how animal mating behavior is understood. The clearest evidence for this centrality of Bateman's Principles comes from how this subject is represented in textbooks. Given that textbooks are designed to represent and teach a given subject or field, those aspects of the field that make it into the textbooks can reasonably be considered the widely accepted foundations of that field, which current practitioners expect future practitioners to accept and build upon. Throughout *Feminism in the Wild*, we authors use the 2019 edition of Dustin Rubenstein and John Alcock's well-regarded *Animal Behavior* textbook as evidence of the discipline's core tenets; we refer to this textbook from here on out as *Animal Behavior*.

The "Sex Differences in Reproductive Behavior" section of *Animal Behavior* opens with a question that takes Victorian-era gender stereotypes entirely for granted. "Why is it more common for males to do the courting, and females to do the choosing?" the textbook asks, without pausing to wonder: do we *really know* that it's "more common," and if not, why do we think it is?[20] *Animal Behavior* also largely dismisses Gowaty and colleagues' work showing fatal flaws in Bateman's 1948 paper: "Although this team was unable to replicate Bateman's results, their findings do not eliminate the important principle that Bateman outlined."[21]

This trajectory of research on Bateman's Principles, and its intertwining with resistance to (and from!) the patriarchy, serves as a clear example of the central claim of feminist science studies: that scientific knowledge is inseparable from the social, cultural, and political contexts in which it is created. Two core concepts from feminist science studies—situated knowledges and standpoint theory—explain this intertwining more fully.

SITUATED KNOWLEDGES AND STANDPOINT THEORY

The idea that who we are affects what we notice about the world is simultaneously unsurprising and complicated. *Of course* we each notice differently—the world is too full, too complex, for any of us to absorb all of it, and so we filter what we experience based on a million different things, including our previous experiences, our interests, and our habits, all of which are shaped by the specific historical moments and cultural contexts of our particular lives. When Ambika notices squirrels, she often compares the squirrels she sees in the US to those she grew up with in India, which are smaller and striped; when Melina sees squirrels, she often considers how they might interact with her dog Pepper.

What we make of our experiences—the feelings, thoughts, and actions they provoke—depends on who we are too. Imagine, for example, how you might respond to a snake, or a dog, if you've been bitten by one before, versus if you've not only *not* been bitten by one but in fact had one as a beloved childhood pet. Imagine, then, how these experiences may seep into the stories you tell about particular kinds of dogs or snakes you've interacted with, or perhaps of dogs and snakes in general. And what if you went on to become an animal behavior scientist: how might your experiences, and the stories you tell about them, seep into the hypotheses you go on to test about the aggressive and social behaviors of dogs and snakes? And now imagine:

What if a vast majority of the other scientists studying dogs and snakes (but not the vast majority of all people) had had roughly the same experiences as you? It seems likely that your research on these animals' aggressive and social behaviors would resonate with most other scientists. Most scientists would agree that the questions you've chosen to ask are interesting, and likely wouldn't bat an eye when your research confirmed that, yes, dogs are indeed quite friendly toward humans, or that snakes are quite aggressive (for example). You'd be suspicious of claims from a minority of scientists, as well as from many non-scientists, that dogs can be quite aggressive and that snakes can be friendly. You might even dismiss their experiences by saying, "well, science tells us otherwise." And you might do all this without reflecting at all on how your early experiences with dogs and snakes shaped your scientific inquiries, because mainstream science does not invite such reflection. In contrast, the field of feminist science studies argues that such reflection is essential.

In the late 1980s, the field of feminist science studies started to gain traction in US academic circles. At that time, following the US civil rights movements of the 1960s and '70s, people of color and white women were making inroads into scientific fields that had historically been the exclusive purview of elite white men. These minoritized scientists were told, however, that in order to be proper scientists and produce neutral, rigorous knowledge, they had to leave their identities at the laboratory door, so to speak.[22] The assumption here was that the mainstream practice of science, established largely by elite white men, was somehow free from cultural background or gendered perspective. Feminist biologist-turned-science studies scholar Donna Haraway called this tendency of mainstream science to presume a superior, politics- and society-free perspective "the god trick of seeing everything from nowhere."[23]

In this political moment, feminist science studies made its central claim, namely that scientists are also people and so cannot help but bring their personal perspectives into their work. And just like all other people, scientists' perspectives stem from their particular life experiences, which, in turn, are embedded in broader historical, political, and cultural contexts. This observation—that how scientists understand and communicate things is just as intertwined with experience, culture, and politics as all other human understandings and communications—is called "situated knowledge," a term coined by Haraway in 1988.[24] The idea that even scientific knowledge is always situated implies that science is not a purely objective, neutral way of knowing, and that what we know about the world through science depends in large part upon *who* gets to do scientific research.[25] Scientists are not a random subset of people, either, and that specificity affects what any of us come to know through science.

To be sure, people working in feminist science studies in the 1980s were not the first to point out that mainstream science is inherently political. For example, Indigenous and Aboriginal practitioners of science, whose own scientific traditions predate and persist beyond colonial contact, have also written extensively since at least the late 1970s about the racism embedded into Euro-American science.[26] And Black Marxist psychiatrist Frantz Fanon (1925–1961) wrote about how the field of psychology justified colonial power dynamics and naturalized racism.[27] These arguments help us see that we need to trace further back to understand science's political entanglements.

The way of knowing that most of us think of as science derives largely from the philosophies and worldviews of elite European men who lived during the Enlightenment (1685–1815).[28] These men were products of their historical time, a time that included the rise of the transatlantic slave trade and European colonial conquests across the globe.[29] Certain social perspectives that were largely taken for granted among elite white men—that men are superior to women, that Christianity is the only true religion, that the "white race" is superior to all others, that animals are objects to be used for human benefit, and that nature is a resource from which commodities must be extracted—were thus also taken for granted by those who got to engage in scientific inquiry, and thereby deeply influenced scientific discourse. Take, for example, this passage from Charles Darwin's *The Descent of Man and Selection in Relation to Sex*, first published in 1871:

> He who has seen a savage in his native land will not feel much shame if forced to acknowledge that the blood of some more humble creature flows in his veins. For my own part I would as soon be descended from that heroic little monkey, who braved his dreaded enemy in order to save the life of his keeper;

or from that old baboon, who, descending from the mountains, carried away in triumph his young comrade from a crowd of astonished dogs—as from a savage who delights to torture his enemies, offers up bloody sacrifices, practices infanticide without remorse, treats his wives like slaves, knows no decency, and is haunted by the grossest superstitions.[30]

Darwin's description of the behavior of "a savage" is consistent with notions of European superiority that were prevalent in Darwin's social and political milieu—one can imagine the "savage" in question describing their own society, not to mention European society, quite differently.[31] But by virtue of being mentioned in a scientific text, subjective and blatantly racist perspectives like Darwin's became part and parcel of the scientific enterprise.

Nonetheless, elite Europeans claimed their science was a superior way of knowing precisely because it supposedly generated only pure, neutral, incontrovertible facts. Because of this claim to objectivity, the entanglement of scientific knowledge with the social, political, and historical contexts of the people producing scientific knowledge has remained largely uninterrogated, not least by scientists themselves. When combined with the elite status of a vast majority of scientists, this uninterrogated entanglement leads to a science that uncritically reflects elite social perspectives. The power that science derives from appearing to be objective in turn serves to reinforce the power held by already-powerful members of society, who are also most likely to be doing the work of scientific research. As a result, and as we show throughout this book, science contributes to the marginalization and oppression of other, non-elite cultures and perspectives, including those of white women, people of color and colonized people, queer and nonbinary people, working class and poor people, and people with disabilities and neurodivergence.

Three additional points are worth noting here. First, the alignment of science with power is not simply a European or Euro-American phenomenon. In India and the Indian diaspora, for example, a majority of scientists (including Ambika) belong to the dominant Brahmin caste of priests and educators who, historically and to this day, hold on to power by appointing themselves both intellectual and moral authorities. They further employ these positions of cultural authority to hold on to economic and political power. The people most marginalized by this rigid, hierarchical caste system were long barred from access to education, and even in 2020 made up less than 5 percent of the scientists at one of India's premier scientific research institutions despite composing close to 70 percent of India's population.[32] Unsurprisingly, science and Brahminism have become intertwined—some Brahmins regard themselves as naturally more curious and patient, and thus better suited to scientific research than members of other castes. One Brahmin scientist justified his vegetarianism (associated with dominant-caste "purity") as "scientific," and cafeterias at scientific research institutions sometimes separate vegetarians from non-vegetarians.[33] In a 1983 paper, ecologist Madhav Gadgil and anthropologist Kailash Malhotra (both members of dominant castes) compared caste categories to biological species and used ecological theories to argue that the centuries-old caste system is rendered especially stable not by the ruthless enforcement of its hierarchies by the dominant castes, but rather because different castes engaging in different tasks in a society reduces competition for resources among them—a clear example of scientists using their science to uncritically naturalize an all-too-human system of oppression.[34] Nonetheless, these associations of science and dominant-caste ideology can be elided because of science's claim to objectivity.[35] This pattern then—of science primarily being the domain of a society's elite and working to naturalize that society's hierarchies while simultaneously denying the political implications of scientific findings—is not unique to the US and Europe.

Second, it can be tempting to think that the easiest way to mitigate the effects of science's historic entanglements with systems of oppression is simply to encourage diversity among scientists on the basis of race or gender, for example. While diversifying the group of people who get to do science is no doubt an important move toward equity and a necessary precursor to diversifying the stories that scientists consider, it is not sufficient on its own. Unless the *practice* of science itself explicitly makes room for scientific inquiry that emerges from different perspectives, scientists will be required either to conform to mainstream stories or to fight an uphill battle against those who insist that the (Euro-American) scientific method is the superior way of knowing the world. Diversity and inclusion are important, in other words, but it's equally important to ask: who gets to invite whom into what existing power structure? As one of Melina's mentors, Black feminist geographer Carolyn Finney, memorably put it: "what are you including me *in*?"[36]

Third, and relatedly, not all perspectives are equally likely to produce novel insights, an idea that is captured by what feminist intellectuals call *standpoint theory*.[37] Black, Marxist, and transnational feminists in particular have observed that the people who are most marginalized and oppressed by a given system are in the best position to identify the faults of that system—their marginalized social locations in systems of oppression, or *standpoints*, give them a unique vantage point into understanding how these systems work (or don't).

Crucially, standpoint theory is not simply about critique. Marginalized standpoints can also reveal different possible solutions to problems—and generate different questions—that members of the dominant group might not even recognize as problems, nor think to ask about. Moreover, standpoint theory is not meant to be a reductive or essentializing form of identity politics, although it has been criticized as such. Not all members of a marginalized group (or any group, for that matter) have the same experiences or think the same things, especially in different social contexts and in different historical periods. Rather, standpoint theory emphasizes *collective* identity as a source of political insight and power. In other words, marginalized standpoints help define broader structural injustices that we can all organize against to our collective benefit, even as marginalized people's individual experiences within an oppressive society inevitably differ in the particulars.

There is ample evidence from the last century or so that as people with different standpoints have become scientists, the content of science has begun to change.[38] The trajectory of research around Bateman's Principles is a clear example: as scientists with explicitly feminist politics (people such as Sarah Blaffer Hrdy, Patricia Gowaty, and Zuleyma Tang-Martínez, among many others[39]) became established in the field, their research made it clear how dominant theories of animal sex reinforced patriarchal values. It's important to emphasize that these scientists did not question prevailing theories of animal sex simply because they were women. Rather, as standpoint theory helps explain, these scientists' insights derived from their political commitments to feminism (no doubt based, at least in part, on their personal but not universal experiences as women), including their willingness to apply explicitly feminist perspectives to the content of their research. As feminist biologist Ruth Hubbard put it in 1988: "I doubt that women as gendered beings have something new or different to contribute to science, but women as political beings do."[40]

Despite all these efforts, still today the political implications and stakes of scientific research are rarely considered explicitly, even when that research is pushing against the mainstream. Bringing feminist analysis into the field of animal behavior science thus requires unearthing the political undercurrents beneath seemingly neutral research. For example, an unassuming paper by behavioral ecologists Hanna Kokko and Johanna Mappes, published in 2013 in a relatively obscure entomological journal, described a way of thinking about female mating behavior that fundamentally undermined Bateman's Principles without being explicitly feminist at all.

"Much of the literature on polyandry asks why females mate [with multiple males]," Kokko and Mappes wrote. "This way of phrasing the issue reveals a curious tendency to think of monandry as a 'null model' such that deviations from it require us to search for a reason. The opposite way to think about the situation is to ask, under what conditions should an individual reject a mating opportunity?"[41] Essentially, Kokko and Mappes asked: is there a story that can be told about females' mating decisions in which mating with multiple males is obviously sensible, rather than something requiring extra explanation? Turns out there is.

The crux of Kokko and Mappes's argument is twofold. First, they imagine that, instead of deciding in advance how many mates they want over their entire lives, female animals make decisions about whether or not to mate with each male they happen to encounter. And second, Kokko and Mappes acknowledge that females don't always get to meet *all* potential mates before choosing whom to mate with. Rather, females encounter potential mates sequentially, and must decide whether or not to mate with them before moving onward to meet others—think less attending a Victorian ball with all eligible bachelors in attendance, from among whom one coyly chooses a husband, and more searching on Tinder, where one must make a decision about swiping right or left before moving on to the next contender.

There is something deeply sensical—or at least, nothing obviously silly—about thinking that female animals make mating decisions on a case-by-case basis. In contrast, imagining that females predecide how many mates they want to end up with seems more unlikely, especially in unpredictable environments, where there's no telling when or even if a female will meet another potential mate. What Kokko and Mappes demonstrated, using a simple mathematical model, was that when a female makes these one-by-one mating decisions as she encounters males at relatively unpredictable intervals, then being choosy and rejecting a mating opportunity (what Darwin, Bateman, and Trivers insisted females must do) is actually too risky. The more males she rejects, the higher her risk of never reproducing at all. Thus, assuming that individual females are universally driven to reproduce (an assumption we will question in later chapters), Kokko and Mappes showed that one should expect females to almost *always* mate with multiple males, because the cost of not mating at all and having a fitness of zero far outweigh the risks (if any) of mating with multiple males. In sum, polyandry is not a conundrum in search of explanation. And while Kokko and Mappes's paper was not explicitly feminist, many decades of feminist analysis on Bateman's Principles preceding this paper made abundantly clear that polyandry was only considered a conundrum because of scientists' sexist assumptions about how female animals move through their worlds and make decisions about whom to mate with.[42] These two efforts—scientific model building and critical feminist analysis—can thus work in conjunction with one another to overturn long-held scientific dogmas based upon limited assumptions rooted in sexism and patriarchy.

Adopting this combined approach requires that scientists get truly comfortable with the concept of situated knowledge. Perhaps without recognizing it, animal behavior scientists *already* have the tools they need to embrace the idea of knowledge being situated and partial. The concept of *Umwelt* offers a complementary way of thinking about situated knowledges in terms that are more familiar to animal behavior scientists, and to those of us accustomed to thinking scientifically about animal behavior. The origin of the Umwelt concept also offers an object lesson in the political complexities that a scientific idea can hold.

THE COMPLICATED CASE OF UEXKÜLL'S UMWELT

In 1909, the German zoologist Jakob von Uexküll coined the term *Umwelt* to describe the worlds that animals perceive through all their senses. Specifically, Uexküll wanted to draw attention to the fact that different animals have very different Umwelten, thanks to their different sensory capabilities: dogs experience detailed smell-scapes through their noses, rattlesnakes visualize thermal landscapes through temperature-sensing pits on their snouts, and seals use their whiskers to perceive the slightest trails of movement through the water. And we humans have *no idea* what it would be like to experience the world in these ways. As philosopher Thomas Nagel famously put it: "I want to know what it is like for a *bat* to be a bat. Yet if I try to imagine this, I am restricted to the resources of my own mind, and those resources are inadequate to the task."[43] Though it remains "beyond our ability to conceive" of the Umwelten of other animals, the last few decades have seen an explosion of research into animals' varied sensory systems.[44] From this research, we humans can do our best to imagine other animals' experiences while understanding that we can never know them fully.

The concept of Umwelt brings to light the limitations of how we humans experience the world. As journalist Ed Yong put it in his 2022 book on the topic, "Our Umwelt is still limited; it just doesn't *feel* that way. To us, it feels all-encompassing. It is all that we know, and so we easily mistake it for all there is *to* know."[45] But on the flip side, being *aware* of our limited Umwelten reminds us that our knowledge is partial and contextual. The concept of situated knowledges offers a very similar reminder—that how we experience the world today depends on who we are and what we have experienced before.

When we start to understand animals' Umwelten, we can begin to make sense of how their own perceptions of the world shape their behaviors. For example, Yong described his own experience of watching a fruit fly *after* learning about fruit flies' capacity to detect gradients of heat in the air they fly through: "I suddenly reconsider the movements of every fly I've seen. Their paths, which always seemed so random and chaotic, now take on an air of purpose, as if the insect is threading its way through an obstacle course of hot and cold that I can't perceive, don't care about, and oafishly wade through."[46] Again, the parallel to the *consequences* of human situated knowledges is clear: how we experience the world in turn influences how we act in it, and we each experience and act differently. Unlike with animals, however, we humans can communicate with one another about our different experiences and actions, which, if anything, ought to make it easier to understand another human's situated knowledge than it is to understand another animal's Umwelt. Animal behavior scientists' relative comfort with the idea of Umwelten affecting behavior thus bodes well for more widespread acceptance of the idea of situated knowledges and its importance to science.

But the specific example of the Umwelten theory also offers us a rather stark lesson in the complexity of reckoning with the political contexts and consequences of scientific ideas. Today, the Umwelten theory might motivate us to find connection across difference; Yong ended *An Immense World* with such an exhortation: "We may never know what it is to be an octopus, but at least we know that octopuses exist, and that their experiences differ from ours. Through patient observation, through the technologies at our disposal, through the scientific method, and, above all, through our curiosity and imagination, we can try to step into their worlds. We must choose to do so, and to have that choice is a gift."[47] However, the scientist who formulated the Umwelten theory to begin with, Jakob von Uexküll, drew entirely different lessons from it.

In addition to being virulently anti-Darwinian (he rejected the idea that all species are related to one another), Uexküll was firmly antidemocratic—so much so that he sympathized with the Nazis—and did not hesitate in wielding his scientific authority to support his political agenda.[48] He used his Umwelt concept to argue against placing decision-making power in the hands of the people, and in favor of something more akin to the Hindu caste system, wherein social hierarchies are naturalized and everyone is seen as serving their predestined purpose within society.[49] Uexküll believed, for example, that the noble work of science ought to be reserved for aristocrats such as himself.[50] Moreover, while Uexküll insisted that all Umwelten are "expressions of the same creative life energy," as historian Anne Harrington put it, he cautioned against "racial mixing" because the "different Umwelten of different racial groups could, only with great difficulty, if at all, ever be reconciled with one another."[51]

It might be tempting to reject the Umwelten concept entirely because of the elitist and racist politics of its creator. But the later use of this same concept in service of building connection among living beings, and indeed our use of it here to illustrate the concept of situated knowledges, suggests that the Umwelt concept can also be applied toward liberatory ends, rather than only toward Uexküll's oppressive ideologies. The lesson we draw from this seeming contradiction is that there are no shortcuts—it is incumbent upon us to carefully think through the political contexts in which ideas were developed, as well as the logics that these ideas naturalize, before deciding how we want to use the idea going forward, and to recognize that we may not all agree on the right course of action. In some cases, such an investigation leads us to set aside whole concepts (Bateman's Principles might be one such concept) and search for different ways to study animal behavior. In the case of Umwelt, we authors elect to hold onto the idea while remembering—and confronting—the vastly different political goals it can be used to support.

Uexküll belonged to an era in which biologists were unabashed about using their scientific research to argue for the validity of their political beliefs; it wasn't until the second half of the twentieth century that animal behavior science acquired a veneer of political neutrality. The next chapter explores this transition in animal behavior science from explicit to implicit political motivation, by first explaining what's known as the *levels-of-selection debate* and then tracing some of its rather chaotic history.

Figure 3.1

Marigold the hen, and her coop-mates, in Oakland, California. Photo: Julia Mitchell.

3 PERSPECTIVES AT WAR

THE HABITS OF HENS

The life of a hen living alone in a cage is very different from the life of a hen living in a cage with many other hens (figure 3.1). On one's own, there's no one to peck at, no one to huddle with if it gets cold, no one to share food with, and no disputes over who gets to settle into the better corners of the cage. So when commercial poultry farmers began to house their hens in groups, they had to contend with different dimensions of these animals' biology. Foremost among these dimensions was the tendency of hens to peck at each other, especially when in captivity and close quarters. In multiple-hen cages, the pecking could get vicious, often leading to injury or death. Some farmers tried to control the pecking by trimming their hens' beaks—no beaks, no pecking. From the hen's perspective, though, beak trimming isn't a great solution; beaks are highly sensitive organs of touch, and hens use their beaks to manipulate objects around them, so beak trimming is not only painful for a hen but also limits her ability to interact with her world.[1] But relative to earlier hen housing methods, multiple-hen cages increased feed efficiency, reduced disease, and lowered production costs.[2] Commercial poultry farmers who wanted to house their hens in multiple-hen cages were caught between a rock and a hard place.

Situations in which plants or animals aren't doing what humans need or want them to do are common in agriculture and animal husbandry. For millennia, farmers and animal breeders have overcome such obstacles with the tool of selective breeding, or artificial selection. To artificially select a population, a breeder identifies those individuals in the population who display the breeder's desired traits, and includes only their offspring in the next generation. Artificial and natural selection are two sides of the same coin. Though one is executed by humans to serve human-desired ends and the other is simply a consequence of natural births and deaths with no predetermined outcome, artificial and natural selection are otherwise the same process. Indeed, Darwin drew much of his inspiration for the theory of natural selection from observing the work of animal breeders.[3]

In the early 1980s, hoping to improve the well-being of hens housed in multiple-hen cages, geneticist Bill Muir wondered whether there was a way to selectively breed these hens to reduce how often they peck at each other. But Muir also knew that for his interventions to be commercially viable, they couldn't come at the cost of reduced egg production. And so Muir designed an artificial selection experiment that began with housing hens in groups of half siblings (that is, all the hens in a group were offspring of the same rooster). In time, these hens laid eggs, and Muir identified which groups of hens produced the most eggs. He then chose offspring from the most productive groups to form the next generation.[4] In other words, rather than selecting the highest-producing *individuals*, he selected the highest-producing *groups* of hens. If you were a hen in this experiment, it wouldn't matter how many eggs you produced—if you were part of a highly productive group, your offspring would make it to the next generation. Think of something like a soccer team, where, at the end of the match, the success of the team depends on the total number of goals the team scores, and not on how many goals each individual player scores. It's the winning team (or, in the case of the hens, their offspring) that advances to the next round of the tournament; the referees do not construct new teams made up of the individual players who scored the most goals.

The logic behind Muir's choice to select groups rather than individuals was both straightforward and unorthodox. When hens are housed in groups, it can be difficult to determine how many eggs any particular individual has laid, which also makes it difficult to pick out which individuals have produced the most eggs. Logistically, it's just easier to select the most productive group. More importantly, however, Muir recognized that the behavior he was actually interested in modifying—how much hens peck at each other—could only be expressed in a group context. It seemed likely that groups of hens who didn't peck much at each other would be in better shape overall, and would produce more eggs. Muir thus hypothesized that selecting *groups* of hens that collectively produced more eggs could reduce those behaviors that harm overall egg production, including pecking at one another (as long as pecking interactions are *inherited*—that is, passed on from groups of mother hens to groups of their daughters; more on the complexity of inheritance in chapter 7). But this hypothesis went against the orthodoxy among animal behavior scientists, which was that selection acts on individuals, not groups. Muir's proposal was thus met with ridicule from his peers. He recalled that when he "suggested that group selection would solve the problem" of hens pecking each other in multiple-hen cages, "few believed me and several laughed openly."[5]

By Muir's own account, his artificial group selection experiment was plagued by various logistical mishaps, but what remained unchanged throughout this nearly decade-long study was the implementation of group selection. The results were striking: after just six generations of group selection (and with no beak trimming after the first generation), hens housed in groups not only survived as long but also produced almost as many eggs as hens housed individually. This change represented a 65 percent increase in egg production in group-living hens, relative to the start of the experiment! Meanwhile, mortality among these hens dropped from 68 percent to 9 percent. Later studies showed that these group-selected hens were less aggressive and lost fewer feathers from stress or pecking than randomly selected counterparts.[6] Muir had addressed the animal welfare problem he had set out to address; he had also shown that artificially selecting groups of animals could lead to individual behavioral change.

In thinking about selection at the group level, it also makes sense to start thinking about fitness not just as an attribute of individuals but also as an attribute of groups. As evolutionary biologist David Sloan Wilson, a longtime proponent of group selection, has written regarding Muir's experiment:

> If by "individual trait" we mean a trait that can be measured in an individual, then egg productivity in hens qualifies. You just count the number of eggs that emerge from the hind end of a hen. [Not simple in practice, but conceivable in principle.] If by "individual trait" we mean the process that resulted in the trait, then egg productivity in hens does not qualify. Instead, it is a social trait that depends not only on the properties of the individual hen but also on the properties of the hen's social environment.[7]

Though the context in which Muir's experiment takes place—multiple-hen cages—is human-made, the conclusions one can draw from the experiment easily extend to the wild. Many animals are social, and *all* animals interact with other members of their species in ways that affect their fitness, implying that an individual's fitness is influenced by their interactions with their group-mates in wild populations as well. Persuasive examples of the influence of social interactions on fitness in the wild come from observations of animals taking care of wounded or otherwise differently abled members of their species. For instance, in their coauthored book *Wild Justice*, animal behavior scientist and cognitive ethologist Marc Bekoff and bioethicist Jessica Pierce told this story of Babyl, an elephant: "Because of an injured rear leg, Babyl could only walk at a snail's pace, and for over a decade and a half, the other elephants in her group have waited for her and fed her. Unescorted, Babyl would easily have fallen prey to a lion."[8] And Babyl's story is not unique; Bekoff and Pierce continued: "There's also the story of a forest elephant who had lost her trunk to a trapper's snare. The injured elephant learned how to drink and how to eat river reeds, the only food she could manage without her trunk. Group members helped to keep their friend alive by altering their own feeding habits, and bringing her reeds. And it has now been reported that all this group eats is river reeds."[9]

Readers may be unsurprised that elephants, who are commonly characterized as one of the more intelligent species, exhibit such evidence of caring and interdependence, but Bekoff and Pierce also recounted several examples of how even so-called lesser animals, like rodents and fish, are frequently observed caring for their companions who move differently due to bodily constraints. And there is, of course, no reason that such social interactions that influence an animal's survival or reproduction should be confined to injured or differently abled individuals.

Thinking of fitness as a socially influenced trait allows one to conceive of natural selection on groups. Just as some individuals survive and reproduce more than others, some groups persist while others do not. If animals' social interactions affect the persistence or extinction of the groups they're part of (and if this effect remains similar from one generation to the next), then the evolution of an animal's social behavior will be shaped by natural selection on groups. Of course, there are tremendous differences between captivity and the wild, not least that the composition of animal groups is potentially far more fluid in the wild, making it much harder for scientists to keep track of which individuals belong in which group or even to know what a "group" is. But the fact that studying groups in the wild might be logistically challenging doesn't render it inconceivable that social interactions influence fitness or that natural selection acts on groups. There is no rule that nature ought to be straightforward to study, or mathematically tractable.

Moreover, neither Muir nor the other contemporary evolutionary biologists who have championed group selection suggest that natural selection acts *only* on groups rather than on individuals or genes. They advocate instead for what's called *multilevel selection*, wherein natural selection acts on genes, individuals, *and* groups. And while it can be intuitive to think about these different forces of selection as acting independently of one another (which would allow us to simply sum up their effects), in fact they *interact* with each other, leading to outcomes that one wouldn't necessarily predict. For example, in decades of experiments (mostly on flour beetles) starting in the 1970s, evolutionary biologist Mike Wade and colleagues showed the following: (1) sometimes group selection and individual selection can cancel each other out, even when they both favor similar changes in the same trait; (2) sometimes group selection is stronger when it acts every few generations than when it acts every generation; (3) sometimes, inconsistent individual selection can enhance the effects of group selection; and (4) often, group selection can produce more substantial and persistent change than individual selection.[10]

It would thus seem that ignoring group selection in our understanding of natural selection is tantamount to deliberately preventing ourselves from fully appreciating the complexity of how animal behavior evolves.

Nonetheless, as recently as 2011, more than one hundred biologists signed a strongly worded statement rejecting group selection as an important evolutionary force in shaping the biology of highly social insects such as ants, wasps, and termites.[11] Their vociferous rejection was a response to yet another attempt by a few evolutionary biologists to argue that group selection is important, and prolonged an already decades-long dispute about whether or not natural selection acts on groups. Indeed, this levels-of-selection debate has remained a fixture in the field of animal behavior science for almost a century.[12] The other side of this debate, the mainstream side that opposes multilevel selection, instead explains the evolution of group-level dynamics with a theory called *kin selection*.

As you read the rest of this chapter, you may notice that we (Ambika and Melina) are taking a side in this debate—we're on the side of the multilevel selectionists, for now. However, our stance does not arise straightforwardly from our socialist politics. If anything, as you read on you will see that biologists can mold whichever theories of natural selection they support to fit a diversity of political perspectives. One can use group selection to argue in favor of fascism or in favor of communism. Individual selection can become a rousing defense of individual liberty and democracy as easily as it supports ruthless capitalism. To us authors, multilevel selection can align at least partially with social anarchism; to Bill Muir, who conducted the hen experiment, multilevel selection supplies an argument in favor of free market capitalism.[13] We're temporarily on the side of the multilevel selectionists largely because, as we explain in the next section, we oppose the kin selectionists' fealty to individual self-interest, their limited understanding of the word "kin," and their insistence that collectives can only be understood by breaking them into smaller parts (a stance known as *reductionism*, which characterizes mainstream animal behavior science and much of contemporary science more generally, and which feminist science studies scholars, among others, have resoundingly criticized[14]). More on all of this in the rest of this chapter; arguments presented in subsequent chapters will release us from the need to take a side in this debate at all. But in the meanwhile, delving into this polarizing levels-of-selection debate helps us all to practice moving through disagreement, and toward holding more complexity, skills that are essential to a feminist approach to science.

GROUPS OR KINSHIPS?

The *Animal Behavior* textbook opens with a classic example of altruism: honey bees that sting you to protect their hive from you, and die shortly after. "Why would an individual kill itself to temporarily hurt you?" the textbook asks, framing this behavior as paradoxical.[15] But altruism is only paradoxical if one insists that natural selection cannot act at the level of the group—it's easy to see how groups might persist longer if some of their members risk their individual benefit for the benefit of the group. Clearly the textbook authors favor a different explanation. "The solution" to this paradox, they write, "lies deep in the portion of the stinging bee's genetic code that it shares with its hive mates."[16] By invoking the bee's genetic code, the textbook is referring to a theory known as kin selection, which allows mainstream animal behavior scientists and evolutionary biologists to square a deep aversion to group selection with the undeniable existence of cooperation ("we both benefit") and altruism ("you benefit at my expense") in animals. Kin selection achieves this by moving *down* a level instead of up, ignoring the group in favor of the gene.

According to the theory of kin selection, one can expect cooperation and altruism *between genetically related individuals* specifically, because of the following (gene-focused) logic. Copies of a gene that are present in two related individuals can pass on to the next generation in two ways: first, when both individuals reproduce individually, and second, when one of the individuals altruistically sacrifices his own survival and reproduction to help his relative increase how many offspring she has. From the perspective of the gene, these two outcomes are indistinguishable—both situations lead to more copies of the gene in the next generation, and so both are equally favored by natural selection acting on genes. In a kin selection framework, altruism among related individuals is actually self-interest at the level of the gene.

What is more, the kin selection framework expects greater self-sacrifice (measured in costs to an animal's own fitness) between individuals when they are more likely to share genes, which in turn implies that individuals will be quicker to sacrifice, and will sacrifice more, for closer relatives compared to more distant relatives or unrelated individuals. Put another way, kin selectionists would expect you to sacrifice the most for an identical twin, equally for your children and your full siblings, less for your half siblings, even less for your cousins, and not at all for an unrelated stranger.

Returning to the hen example: a kin selectionist would argue that Muir's hens in multiple-hen cages evolved to peck less at one another not because of group selection but because the hens in the group were half siblings. Pecking less at your half sibling might mean that you lose the competition for food and lay fewer eggs—*you* are making a sacrifice. But the *genes* that you share with your half sibling are doing no such thing if, as a result of the reduced pecking, your half sibling lays more eggs. If groups are made up of relatives, the kin selectionists claim, there is no reason to think at the level of the group at all.

But to ignore the group is to ignore the fact that kin selection can only explain dimensions of social behavior and cooperation that can be readily broken into separate parts and added up. Instead of assessing the persistence of groups as a whole, kin selectionists try to tally all the costs and benefits of all social interactions to each individual in the group, weighted by how closely related the interacting individuals are. Imagine you're a hen in a group of seventeen hens. Instead of considering the collective thriving of your group, a kin selectionist would focus on how the group is made up of six half siblings, four full siblings, and seven individuals who are unrelated to you but related to each other; you peck less at the three individuals perched closest to you but peck more at the five individuals who you don't interact with that often, and team up with an unrelated individual to peck at the most aggressive hen in the group, who happens to be your full sibling. The kin selectionist would try to apportion the costs and benefits of your pecking behavior to each individual in your group. Such an approach can seem precise and perhaps logistically tractable, but in fact can *never* capture the synergistic effects of relationships among individuals where the whole is more (or less, or simply different) than the sum of its parts.[17]

Proponents of multilevel selection have long described how kin selection transforms all relationships into self-serving ones. For example, David Sloan Wilson has described being frustrated by the kin selectionists' insistence on "transmuting altruism into selfishness."[18] According to their individualistic framings, he wrote, "A relative helping another relative became an individual helping its genes in the body of another individual," and acts of altruism toward nonrelatives became nothing more than "a matter of scratching your back so that you'll scratch mine."[19] In this way, altruistic behaviors "collapse into a heap of self-interest" instead of being interpreted as benefiting the group.[20] Such a view is not politically neutral. Indeed, the logic of kin selection is famously captured by a quote from the Conservative British prime minister Margaret Thatcher: "There is no such thing as society. There are individual men and women, and there are families."[21]

Denying the existence of society makes it easy to ignore the existence of relationships among individuals and among families, relationships that lend structure to a society and shape the experiences of individuals in this society, including how resources flow among individuals. And of course, not all of these socially and materially important relationships exist between mates or genetic relatives. Stepping back to think of human relationships more expansively, one can turn to decades, if not centuries, of anthropological and archaeological research that has led to a "hard-won understanding that human kinship is not a naturally given set of 'blood relationships' but a culturally variable system of meaningful categories."[22] Whether or not a person calls someone a "brother" or an "uncle" can, depending on their culture, have much more to do with how much time they spend together, how they share resources, and where they live relative to one another, and much less to do with their genetic relatedness. This highlights how the kin selectionists' use of the word "kin" to mean "genetic relatives" is highly *ethnocentric*, derived from Western cultural norms alone and erasing numerous other kinship arrangements in humans and animals alike.[23] As cultural anthropologist Marshall Sahlins put it: "Of course it is true that all Americans are human, but it is not true that all humans are American—and still less that all animals are Americans."[24]

Social and cultural relationships can also determine how individual animals relate to one another. Consider the lance-tailed manakin, a Central

American bird species in which males pair up to sing and dance together. They trill and chirp persistently while leaping elegantly from slim branches into the air in synchronized circles, one ascending as the other descends. The males' striking blue shoulders and red caps contrast sharply against the black feathers covering the rest of their bodies. Together, these colors, sounds, and movements create quite a spectacle in the forest understory. An olive-green female watches them, and as she moves closer and further, the two male birds change tempo and position. Sometimes, but not always, the female mates with one—only one—of the dancing males.

These singing and dancing duets were the subject of behavioral ecologist Emily DuVal's early research. DuVal wondered why the second male stuck around spending his energy on all the dancing and singing if he never got to mate and therefore couldn't directly increase his fitness. It appeared that the "beta" male was behaving altruistically toward the "alpha" male, and DuVal wanted to understand why. When she began this research, DuVal "really thought relatedness made the world go round."[25] She expected that the pairs of males would be closely related to one another, which would imply that, according to at least some of their genes, the beta male's sacrifice of time and energy in support of the alpha male's fitness was not really a sacrifice at all. But instead, her data revealed that males in these pairs were no more closely related to one another than one would expect by chance.[26] Cooperation among male manakins—individuals pairing up, singing and dancing, practicing, and performing together over the course of months, if not years—is thus a social, maybe even cultural, interaction that is not determined by genetic relatedness. Yet, given the diversity across human cultures in how kin relationships are conceived, there are undoubtedly people who might watch the pairs of unrelated male manakins singing and dancing together for years on end, and consider them brothers.

Even within Euro-American society, kin selection reflects the gendered, racial, and classed positions of its proponents, who somehow find it inconceivable, based on their lived experiences perhaps, that someone might care for another regardless of their degree of genetic relatedness. In contrast, marginalized communities within Western societies create their own meanings of kinship—and altruistic social relations—precisely in response to an oppressive dominant culture. For example, people living with disabilities often build networks of mutual aid, particularly during times of crisis when authorities tend not to adequately help people with particular needs (such as power outages that may be especially dangerous for people who depend on electrical machines to breathe, or evacuation routes that are not wheelchair accessible).[27] In the US context, African Americans have created alternative traditions of kinship because of the ways in which enslavement tore biologically related families apart, a violent history that is simultaneously bound up with economic and property relations.[28] And the queer practice of creating chosen family offers yet another example of kinship that isn't necessarily rooted in biological relationships. Because queer and trans folks are so frequently ostracized, or worse, from the families and communities into which they are born, they have embraced the process of convening their own "chosen" (and multispecies!) families, intentionally claiming kinships that extend beyond a conventional family structure.[29]

Thus, there are a multitude of standpoints from which kin selection's focus on genetic relatedness above all else seems an awfully narrow-minded and limiting approach to understanding animal cooperation and altruism. Black feminist writer Alexis Pauline Gumbs responded poignantly to a group of scientists who expressed surprise over a female dolphin who had adopted and lactated for an unrelated calf (whose gestational mother had died):

> Why, as the authors of the study wonder, do mammals do this, when, as they put it, "the behavior is so costly?" To them, "it is unclear why an animal would invest its resources in this manner." Does it seem unclear to you? Who here has not been mothered by someone genetically and socially distant from your birth situation, at some necessary time? And if you have ever shared something, taught someone, shared responsibility for someone's wellness for even a part of their journey, how would you measure what you gained from that potentially "costly behavior"? We call it love.[30]

Over the course of history and across the globe, the relationships that structure human and animal societies have included all manner of complicated tensions between self-interest and generosity. But the specifics of this diversity are often lost in humans' desires to come up with grand, general theories of what people and other animals are like. As cultural anthropologist and anarchist David Graeber and archaeologist David Wengrow put it in their 2021 book *The Dawn of Everything*: "one of the most pernicious aspects of standard world-historical narratives is precisely that they dry everything up, reduce people to cardboard stereotypes, simplify the issues (are we inherently selfish and violent, or innately kind and cooperative?) in ways that themselves undermine, possibly even destroy, our sense of human possibility." Moving past these stereotypes is going to be essential for addressing some of the key political questions of our time—how do we hold both individual freedom and collective well-being as political values? Do systems change (and *how* do they change) because of the actions of individuals or the actions of groups?

At present, the levels-of-selection debate remains highly skewed in favor of an approach to natural selection that takes self-interest and competition for granted, similarly constraining our sense of possibility in the realm of animal behavior. By rendering selfishness an indelible part of how nature works, mainstream science makes it all the more challenging to escape the belief that animals in general, and humans in particular, are inherently self-interested. A brief recounting of the history of the levels-of-selection debate, offered in the next section, illuminates how mainstream animal behavior science became so thoroughly individualist.

THE RISE OF INDIVIDUALISM

In the first half of the twentieth century, biologists broadly agreed that scientific inquiry about competition and cooperation in animals *ought* to inform human politics and ethics. The levels-of-selection debate was therefore unabashedly political back then, and the idea that natural selection could act on groups was neither mainstream nor summarily dismissed.[31]

In his 1902 book titled *Mutual Aid*, Russian polymath and anarchist revolutionary Peter Kropotkin claimed that "sociability is as much a law of nature as mutual struggle." Before beginning his own observations in Siberia, Kropotkin had been impressed by Darwin's *On the Origin of Species*, and consequently expected to see competition everywhere. However, "facts of real competition and struggle came very seldom under my notice, though I eagerly searched for them."[32] Instead he described examples of animals—from burying beetles to crabs to eagles to reindeer—supporting one another, particularly in challenging environmental conditions. Kropotkin's observations were deeply and explicitly intertwined with his collectivist politics and ethics. As he put it: "In the practice of mutual aid, which we can retrace to the earliest beginnings of evolution, we can thus find the positive and undoubted origin of our ethical conceptions."[33] In the 1920s and '30s, American biologist, committed Quaker, and anti-war activist Warder Clyde Allee was inspired by Kropotkin's findings, and demonstrated similar results in the lab. For example, he placed goldfish into toxic solutions of colloidal silver and showed that groups of goldfish were able to survive longer in these extremely harsh conditions than individual fish.[34]

Allee was one of several biologists at the University of Chicago who, to varying degrees and with various political motivations, came down strongly in favor of animal biology being shaped by cooperation and group selection rather than competition and individual selection.[35] Others included Alfred Edwards Emerson, who "championed [the] concept of the superorganism to legitimate his own political views about democratic society."[36] Yet still others disagreed with the University of Chicago biologists' "'aggregation' ethics."[37] Paleontologist George Gaylord Simpson, for example, saw any consideration of collective welfare as a concession to totalitarianism, and espoused extreme individualism instead, in both human politics and organismal evolution. In 1941 Simpson wrote: "The group is a collectivity of individuals. It has no entity except as derived from the relationships of individuals. It does not evolve except as individuals prosper."[38]

This debate between the individual selectionists and group selectionists was playing out in the broader economic and political context of the Great Depression, the rise of totalitarianism across the globe, and the Second World War. Far from denying or separating themselves from this context, biologists studying animal competition and cooperation in the early twentieth century tended to focus their research on those theories that lent support to their political views. As historian Gregg Mitman put it, biologists of this era believed that "the task of the biologist was to discover nature's moral prescriptions and thereby serve as savior of society."

In the second half of the twentieth century, the mantle of group selection, as well as that of society's savior, was taken up by British zoologist Vero Copner Wynne-Edwards, when he noticed that populations of animals do not appear to steadily increase in number (as the global human population does). Darwin had taken this same observation to mean that many individuals must die when animal populations get to be too big. Wynne-Edwards, however, wasn't so sure—he didn't think that forces such as predation, starvation, and disease acted frequently enough to keep populations in check. Instead, he hypothesized that groups of animals regulate their own population growth, to their collective benefit, through social behaviors that prevent too many individuals from being born in the first place, and that such mechanisms of population regulation must have arisen from natural selection acting on groups rather than individuals.[39]

Wynne-Edwards's appreciation of animals' capacity to seemingly regulate their own population growth stood in stark contrast to his fear of humans' inability to do the same.[40] His panic about overpopulation had a distinctly racist dimension to it. He conceded that "for 5,000 or 10,000 years the advanced peoples of the Western world and Asia have increased without appearing to harm the world about them or endanger its productivity."[41] Nonetheless, he saw fit to issue warnings about our future now that there might be people (implied to be less advanced) in other parts of the world whose populations were beginning to grow too, and needed to be controlled. Wynne-Edwards's perspective was not all that far removed from ecofascism, which contemporary journalist Naomi Klein has defined, powerfully and succinctly, as "environmentalism through genocide."[42] Moreover, he wasn't the only scientist advocating for population control at the time. Far more famously, ecologist Paul Ehrlich's 1968 book *The Population Bomb* posed this stark, fear-mongering question in its very subtitle: *Population Control or the Race to Oblivion?* (Ehrlich's book sparked horrific programs of forced sterilization across Asia, Africa, and Latin America that continued well into the twentieth and twenty-first centuries.)[43]

But the backlash to group selection, which began soon after Wynne-Edwards's rise to prominence, had little to do with his thinly veiled racism or ecofascism and much more to do with his focus on the success of the collective at the expense of the individual. Political context no doubt shaped this backlash to group selection. As historian Ayelet Shavit described, "In the midst of the Cold War, cooperation for the benefit of the whole was associated with group conformity and Communism, while diversity and conflicts of interest seemed to ensure the existence of democratic groups."[44]

Evolutionary biologist George Williams's 1966 book *Adaptation and Natural Selection* roundly dismissed the possibility that natural selection could lead to the evolution of individual self-sacrifice for the common good.[45] He argued instead that individual selection would always dominate over group selection, and that a vast majority of adaptations could be explained by selection at the level of individuals, not groups. For Williams, opposing Wynne-Edwards's take on group selection with an extremely individualistic approach was an attempt to safeguard against the co-optation of the theory of natural selection for totalitarian ends. He reminded his readers that selection at the group level is not all about cooperation—it also requires some groups to succeed at the expense of others. "To claim that [selection at the group level] is morally superior to natural selection at the level of competing individuals would imply, in its human application, that systematic genocide is morally superior to random murder," he wrote.[46] Evolutionary biologist Bill Hamilton, who first came up with the theory of kin selection, similarly saw a direct throughline from advocating for group selection to spreading fascist and communist propaganda.[47]

Williams's strongly individualist approach to understanding natural selection and Hamilton's gene-centric approach to explaining altruistic behaviors via kin selection were brought together in the 1975 book *Sociobiology* by ant biologist and behavioral ecologist Edward O. Wilson. *Sociobiology* presented a sweeping framework for explaining the behavior of all animals, including humans. It aimed to subsume whole disciplines of the social sciences and humanities into evolutionary biology, and to explain vast swathes of human culture as adaptations driven by natural selection at the level of the gene. "In a Darwinist sense the organism does not live for itself . . . it reproduces genes, and it serves as their temporary carrier," E. O. Wilson wrote, on the book's very first page.[48]

In a way, the sociobiologists were also proponents of multilevel selection, but the levels they focused on did *not* include groups, only genes and individuals. They would readily switch back and forth between describing selection as acting on these two levels, and yet the conflation was not seen as contradictory because individuals themselves were thought to exist solely for the purpose of replicating the genes within them. For example, in his bestselling account of the sociobiological point of view titled *The Selfish Gene*, behavioral ecologist Richard Dawkins declared that "it is often tedious and unnecessary to keep dragging genes in when we discuss the behavior of survival machines [Dawkins's term for individual organisms]. In practice it is usually convenient, as an approximation, to regard the individual body as an agent 'trying' to increase the number of all of its genes in future generations."[49]

This easy switching between individual and gene allowed Dawkins and other sociobiologists not only to fervently insist on individual self-interest as a fundamental component of natural selection, but also to make selfishness seemingly immutable, by describing it as a feature of organisms' genes (more on this seeming immutability in chapter 7). An individualist approach to natural selection prompts one to expect that most of the relationships between individual animals will be competitive, self-interested, and aggressive.[50] Dawkins bluntly wrote, for instance: "If you look at the way natural selection works, it seems to follow that anything that has evolved by natural selection should be selfish."[51] Or as evolutionary biologist Michael Ghiselin put it, even more dramatically: "Scratch an 'altruist' and watch a 'hypocrite' bleed."[52]

The rise of sociobiology in the 1970s was rapid. Williams himself was surprised by how quickly and completely his ideas became accepted in the field. As he reflected in 1996: "I fully expected that the perspective I urged would ultimately be accepted as orthodox; however, I did not at that time expect it to prevail so soon."[53] It quickly became unthinkable for animal behavior scientists to consider any frameworks other than those of individual self-interest for understanding animal behavior. Witness, for example, behavioral ecologist Bernd Heinrich's 1989 musings on how ravens interact with one another while foraging in the winter:

> the birds, which had always seemed to me solitary animals, [were] doing something solitary animals are not "supposed" to do: They were sharing valuable food—those who had, it seemed, were giving to those who needed. It was the most left-wing behavior I had ever heard of in a natural system. Furthermore, it did not make sense. (As a biologist interested in how things work, I always look for some evolutionary, self-serving reason for why animals do things, although this is totally apart from the animals' *motives*, and even more removed from what "ought" to be in terms of human behavior.) This time my mind failed to provide a clearly selfish, evolutionary cause for the apparent sharing, and that failure gave me an instant adrenaline rush. I felt that I might not only learn something about ravens, but also something of larger theoretical value. . . . The question almost lunged out: Why were the ravens sharing? What is the underlying pattern that explains the anomaly?[54]

Notice Heinrich's insistence that the fundamental definition of a biologist is one who looks for "some evolutionary, self-serving reason" for animal behavior. And while his mind's failure to "provide a clearly selfish, evolutionary cause" gave him a momentary rush, it did not spur him to question or shift his belief in the natural law of selfishness. Rather, Heinrich concluded that animals sharing with one another—"left-wing behavior"—can only be an anomaly.

In the next chapter, we build a deeper understanding of the political stakes of mainstream animal behavior science's turn toward individualism. To do so, we examine which behaviors scientists come to expect from animals, and which behaviors they consider to be anomalies. The concept of the *scientific paradigm*, and feminist critiques of this concept, turn out to be crucial for making sense of these expectations and anomalies. We begin our journey toward scientific paradigms by considering the behavior of a certain rabbit named Bun, and the complex associations between his behaviors and capitalism.

Figure 4.1

Cinnabun, a rabbit, foraging on greens in an apartment in Cambridge, Massachusetts. Photo: Ambika Kamath.

4 HIDING IN PLAIN SIGHT

A TALE OF AN IRRATIONAL RABBIT

When Ambika was in graduate school, she had the remarkable good fortune to tend to her friend's pet rabbit, Cinnabun, or Bun, for short, over several Christmas breaks (figure 4.1). Tending to Bun didn't involve anything complicated—just cleaning out his cage, replacing his hay, feeding him some fresh vegetables, and keeping him company for a while, possibly with some ear scratching and gentle petting. Worthwhile, emotionally satisfying labor, but not intellectually captivating to an animal behavior scientist fixated on the question of *why* animals behave in the ways they do. And so Ambika decided to test out some classic animal behavior theory on her temporary rabbit companion.

Bun lived in a small rabbit pen in a Massachusetts living room, with no other rabbit residents. The only interesting behavior Ambika could reliably hope to see Bun engage in was eating. The obvious theory for Ambika to explore was the *only* theory that animal behavior scientists consider when trying to understand how animals search for and consume food: optimal foraging theory. Under optimal foraging theory, animals are (unsurprisingly) expected to behave *optimally*, meaning that individual animals are expected to maximize the "returns on their investments" by expending as little energy as possible and gaining as much energy as possible while foraging. This optimization—of the difference between energy obtained and energy expended—is expected to shape how animals search for and choose from among many different types of foods. In short, optimal foraging theory expects animals to be *efficient* foragers.

Ambika arranged piles of vegetables for Bun. The piles were of different sizes, and were placed at different distances from where Bun sat, rather placidly, on the kitchen floor. Some piles were made of dandelion greens, while others had lettuce or mint. There might even have been a pile or two of carrots. She hoped to see anything resembling efficiency in the rabbit's choice of foraging pile—choosing larger, closer vegetable piles over smaller, more distant ones, for example, or more sugary carrots over more watery lettuce. But Bun did not comply—there was no sense to be made of his haphazard nibbling.

One can think of countless reasons why, on a chilly December afternoon, a solitary pet rabbit shuffling across a kitchen floor would not behave exactly as an animal behavior scientist expects. Bun was well-fed, faced no competition from other rabbits, and, on most days, wasn't expending a lot of energy. Perhaps his behavior was exactly optimal for these specific circumstances. But demonstrating optimality would seem to require exploring all the possible influences on Bun's foraging behavior, which would certainly take longer than a few Christmas breaks. And ultimately, the environment—about as human-made as one might imagine—likely had a tremendous influence on Bun's behavior, but its effects couldn't really be accounted for. One might argue that optimal foraging theory is best tested on animals in their natural habitats.

It turns out that even in natural settings animals often don't behave in ways that an optimal foraging model would predict. For example, gray squirrels eat more hickory nuts than is optimal; horned lizards do not spend an optimal amount of time in each patch of ants they feed on; and honey bees seem to disregard how dense flowers are in each patch, how much nectar different flowers produce, and how long it takes to access nectar in different flowers while feeding, leading to decidedly nonoptimal foraging behavior.[1] Yet again, animals may be found ignoring our expectations of them.

Confronted with such unexpected results, animal behavior scientists start modifying their models. Perhaps if the models account for animals' imperfect knowledge of the conditions, or environmental complexity, or the presence of predators, then researchers will find that these animals are behaving optimally after all. If one changes the length of time that one deems relevant to foraging, then one can show that horned lizards behave in a manner consistent with optimal foraging theory. If one accounts for variation in nutrient composition across different kinds of nuts, then the data seem to support the conclusion that gray squirrels forage optimally. And if one thinks about how individual honey bees are different from one another, then one might be able to deduce that they each forage optimally after all.[2] As the *Animal Behavior* textbook puts it: "If the predictions of an optimality model are based on faulty assumptions and fail to match reality, researchers will reject that model. This does not mean the [optimality] approach is wrong, however. It may simply mean that a better model is needed."[3]

The crux of the problem with optimal foraging theory is that its core assumption—the assumption of optimality itself—is questioned only rarely. One such rare occasion took place in a 1987 paper, rather bluntly titled "Eight Reasons Why Optimal Foraging Theory Is a Complete Waste of Time." In this paper, biologists G. J. Pierce and J. G. Ollason argued that the assumption of optimal behavior in animals is not only unwarranted but also entirely untestable. When studying an animal whose behavior doesn't match an optimal foraging model, it is *always* possible for a researcher to add more biologically plausible factors into the model, and repeat this process until the "better" model's predictions match the animal's behavior. And when an optimal foraging model is infinitely malleable in this way, there are *no* results from the model that would allow a scientist to conclude that an animal does *not* forage optimally. Bafflingly, this substantial limitation is both broadly recognized and almost entirely unheeded by animal behavior scientists. As Piece and Ollason wrote: "Although most students of foraging behavior admit that the assumption of optimality cannot be tested, it seems to be forgotten that this means there can be no evidence for optimal foraging."[4]

Microeconomists hold exactly the same assumption of optimality when trying to make sense of human behavior—whereas animal behavior scientists assume that animals will optimize their energetic returns while foraging, microeconomists assume that we humans will optimize our monetary returns on investments when making financial decisions. In philosophical terms, individuals who behave in this optimal fashion are "rational agents" whose behavior is described by rational choice theory, a framework that can be traced to eighteenth-century Scottish philosopher and "father of capitalism" Adam Smith. Armed with the assumption of rational agency, both microeconomists and animal behavior scientists use the mathematical tools of game theory to understand and predict the behavior of rational agents.[5]

But the core assumption of optimality (or rationality) propounded by rational choice theory broadly fails in humans too: over and over again, humans are found to make *non*optimal, *ir*rational choices.[6] The field of behavioral economics has tried to remedy these mismatches in much the same way that animal behavior science approaches animals' nonoptimal foraging behavior: by adding caveats and mitigating factors that try to demonstrate how these creatures are actually behaving optimally after all. Historian Robert Skidelsky described this tendency with frustration in *What's Wrong with Economics?*, writing that "these concessions to reality produce incoherence, not progress."[7]

Ecologists and evolutionary biologists doing the early work on optimal foraging theory were transparent about the parallel logics of capitalist economics and animal behavior science. For example, Robert MacArthur and Eric Pianka open their 1966 paper "On Optimal Use of a Patchy Environment" with the sentence: "There is a close parallel between the development of theories in economics and population biology."[8] This early work on optimal foraging theory was part of the turn toward the sociobiological approach to animal behavior science, and the rise of sociobiology coincided with the wider spread of a specifically neoliberal form of capitalist economics, politics, and policymaking in the US, UK, and beyond.[9] It is thus not surprising that sociobiology and neoliberal capitalism depend on identical assumptions about how individuals, whether human or animal, behave: individualistically and efficiently.

The assumption that humans behave like rational agents is central to how capitalism is conceived of. As Marxist philosopher Søren Mau explained: "the agents who engage in transactions on the market are assumed to be isolated, hyper-rational, utility maximising individuals with infinite and infallible information," and once humans are assumed to *always* be such agents, regardless of social and political context, then "the need to explain the existence of capitalism conveniently disappears: the capitalist economy appears simply as what happens if human nature is allowed to unfold without impediments."[10] In other words, capitalism is nothing more than what emerges when rational agents behave rationally.

The assumption that animals behave like rational agents is central to how mainstream animal behavior science is conceived of. Optimality thinking is not restricted to scientific research on foraging—it extends to the *entirety* of how scientists study animal behavior. In the sociobiological tradition described in the last chapter, which continues to dominate mainstream animal behavior science today, animals are not only thought to forage optimally but also to process information as efficiently as possible, to make optimal mate choices, and to behave as though they are performing complex cost-benefit analyses prior to and during every social interaction. The specific quantity being optimized may differ across different contexts. Whereas animals are expected to maximize energy returns when foraging optimally, they're expected to maximize their odds of survival while evading a predator, and to find some optimal combination of offspring quantity and quality during reproduction. In any case, all these optimized metrics funnel into the overarching metric of *fitness*—every dimension of an animal's behavior is expected to ultimately optimize its survival and reproduction. Even groups of animals can be assumed to behave optimally together, optimizing the fitness of the collective rather than the individual.[11]

Again, parallels with capitalism are instructive. In the world of capitalist economics, individuals (or firms) do not just maximize returns on financial investments. Their goals are broader: they seek to maximize *utility*, a notoriously slippery concept that boils down to whatever people value.[12] Similarly, in mainstream animal behavior science, individual animals (or groups of animals) are expected to maximize more than just their energetic returns while foraging, for example; they are expected to maximize their *fitness*, another notoriously slippery concept that loosely captures an animal's success at surviving and reproducing.[13]

The near-exact alignment of the individualist approach of sociobiology and the individualist ideology of neoliberal capitalist economics meant that the rise of one supported the rise of the other.[14] As sociologist and science studies scholar Eileen Crist argued, the individualist and optimality-focused outlooks of sociobiology became so prominent *precisely because* economic frameworks of costs and benefits, competition, and success were widely and uncritically deployed across much of Euro-American society. Sociobiology, in turn, naturalized the logic of neoliberalism. An "economic style of reasoning"[15] thus captured science and society alike, and came to undergird descriptions of how both humans and animals behave. Its values came to be taken for granted as simply how nature works, or ought to.

Of course this "economic style of reasoning" is not apolitical. Sociologist Elizabeth Popp Berman has written about how, in the realms of government and policy, such an economic style of reasoning is "perceived as politically neutral but . . . nevertheless contains values of its own—values like choice, competition, and, especially, efficiency."[16] It "portrays itself merely as a technical means of decision-making that can be used with equal effectiveness by people with any political values. This, though, is a ruse: efficiency is a value of its own."[17]

Early adopters of optimality thinking in animal behavior science seemed to recognize that they were making a substantial assumption by thinking of natural selection as an efficient, optimizing force. For example, MacArthur and Pianka wrote, in their 1966 paper: "Hopefully, natural selection will often have achieved such optimal allocation of time and energy expenditures, but such 'optimum theories' are hypotheses for testing rather than anything certain."[18] Of course, as we saw previously, the assumption of optimality didn't end up being tested, because it is not testable. There are no data that can lead a scientist to conclude that an animal is *not* behaving optimally, because one can always imagine another fitness cost or benefit of the behavior for which scientists have not yet accounted.

Questioning the assumption of optimality is not the same as questioning whether the process of evolution by natural selection happens at all. Marxist biologists Richard Levins and Richard Lewontin made this point when clarifying the three conditions that must be met for a population to evolve due to natural selection.[19] First, individuals in the population must vary in their morphology, physiology, or behavior. Maybe different individuals run at different speeds, or eat different foods, or have different temperature preferences, or vary in color. Second, this variation must be inherited by these individuals' offspring. The offspring of faster parents must be, on average, faster than the offspring of slower parents (and if they aren't, then speed will *not* evolve over time as a result of natural selection). And finally, this variation among individuals must be associated with variation in fitness. Faster individuals must be more likely to survive and have more offspring than slower individuals. But nothing in these three conditions of natural selection suggests that *only* the most optimal animals will prevail. It isn't *only* the fastest gazelle who manages not to be eaten by a cheetah and who gets to reproduce. Rather, all the gazelles who run fast *enough* to not be eaten by cheetahs may survive and produce offspring. As cultural anthropologist Marshall Sahlins described in his 1977 book titled *The Use and Abuse of Biology* (an immediate and direct rebuttal to E. O. Wilson's *Sociobiology*): "Selection is not *in principle* the maximization of individual fitness but any relative advantage whatsoever. . . . It is thus important to note that while selection may specify a direction of change, it does not specify the final outcome . . . traits representing greater genetic fitness will spread in the population, but this does not of selective necessity entail that these traits will continue to improve or be perfected to the point of structural and functional optimization."[20] Or, a bit more simply: the logic of natural selection does not imply the "survival of the fittest"; it implies only the "survival of the fit enough."[21]

Though the process of natural selection proposed by Darwin can be separated from optimality thinking, Darwin's own ideas about the theory of natural selection were certainly influenced by contemporaneous economic and political thought. He was inspired in large part by clergyman and political economist Thomas Malthus, who claimed that, because the "passion between the sexes" was so strong, the growth of human populations would always outpace increases in food supply, necessarily leading to a "struggle for existence" among individual humans.[22] Darwin extended the same argument to other organisms as well. Malthus's views in turn echoed seventeenth-century philosopher Thomas Hobbes's characterization of human life under the "state of nature" as "solitary, poor, nasty, brutish, and short"—in other words, rife with struggle, competition, and self-interest.[23] Moreover, Darwin wrote *On the Origin of Species* less than a century after the economist Adam Smith wrote *The Wealth of Nations*; Smith described how the "invisible hand" of the market could produce successful societies in much the same way that Darwin later conceived of the invisible hand of natural selection producing well-adapted populations.[24] The logic of adaptation by natural selection was simply, according to philosopher Bertrand Russell, among others, "an extension to the animal and vegetable world of laissez faire economics."[25] Authors of *The Communist Manifesto* Karl Marx and Friedrich Engels corresponded with one another about *On the Origin of Species* shortly after its publication, with Marx commenting: "it is remarkable how Darwin recognizes among beasts and plants his English society." Later, Engels described this transmutation of capitalism into adaptation by natural selection as a "conjurer's trick": "the same theories are transferred back again from organic nature into history and now it is claimed that their validity as eternal laws of human society has been proved."[26] Given these intertwined origins of adaptation by natural selection and capitalism, it is not surprising that, during the twentieth century, as economic theories rooted in capitalism became more focused on the optimal behavior of individual humans, biological theories rooted in adaptation by natural selection became more focused on the optimal behavior of individual animals.

Our critique of the use of optimality thinking in understanding animal behavior does not mean that animals (or the process of natural selection) *never* optimize. It is entirely possible that some animal communities are identical to an idealized capitalist society made up of perfectly rational agents. It would be foolish for us to claim that *none* of the millions of animal species out there face relentless resource scarcity and respond to it with ruthless self-interest, especially given the increasingly challenging environmental conditions produced by human-caused global warming and habitat destruction. (Although Kropotkin's observations of animals engaging in mutual aid in the harsh conditions of Siberian winter suggest that even under scarcity, self-interest isn't always a given.) But it is equally, if not more, foolish to assume that *every* animal—indeed, every organism—lives its life by the economic rules that happen to dominate human society right now. We humans are free to look at the world through the lens of optimization, and doing so might be illuminating, especially if it is one of many lenses we look through. However, we only limit our understanding of nature when we assume that nature must always be already optimal or always in the process of optimizing.

Nevertheless, the logic of optimality thinking—"survival of the fittest" logic—persists. Historian and philosopher of science Thomas Kuhn's concept of the *scientific paradigm*, and feminist critiques of this concept, help us understand the persistence of optimality thinking.

PARADIGMS: WHAT AND WHY

When animal behavior scientists ask *why* an animal behaves a certain way, they're in search of a story that demonstrates how the behavior is optimal. Explanations in animal behavior science are thus always somewhat of a foregone conclusion. The answer to *why* a behavior exists is *always* expected to be "because it optimizes fitness." The details of the specific costs and benefits need to be worked out, and doing so might yield interesting insights into some particular animal's biology. But the broad shape of the explanation is predetermined, precisely because the assumption of optimality underlying such explanations is never questioned.

This description of how optimality thinking works in mainstream animal behavior science is an example of what Kuhn called *normal science.* These periods of normal science follow the birth of *scientific paradigms*, which Kuhn defined as "universally recognized scientific achievements that for a time provide model problems and solutions to a community of practitioners."[27] During a period of normal science, scientific inquiry proceeds according to the bounds delineated by the paradigm—only questions and answers that make sense within the paradigm are considered reasonable. Indeed, outcomes of research questions in periods of normal science "can be anticipated, often in detail," and the work of research within a paradigm thus comes to resemble "puzzle solving."[28] When scientists are working within a paradigm, they know that they *will* be able to solve the puzzle, and they know roughly what the solved puzzle is supposed to look like even before they begin solving it.[29] Kuhn's description of normal science sounds an awful lot like optimality thinking within mainstream animal behavior science—optimality thinking is a paradigm.[30]

In Kuhn's analysis, paradigms quickly become untestable. Pieces of data that don't seem to match the expectations of the paradigm are simply absorbed into it. Kuhn called such pieces of data "anomalies." Within a paradigm, anomalies are explored until "the anomalous has become the expected."[31] What once seemed to be a new fact comes to resemble the old facts. Challenges to a paradigm are initially met with resistance, but eventually become assimilated into the paradigm. In fact, we can recognize a paradigm when we see this fate of anomalies; in a period of normal science, Kuhn says, scientists "will devise numerous articulations and ad hoc modifications of their theory in order to eliminate any apparent conflict."[32]

By now, this response to anomalies ought to sound familiar. Female anole lizards and fruit flies that mate with multiple males, hens that don't peck at each other very much, bees that die after stinging, gray squirrels that feed on too many hickory nuts—all of these animals' behaviors are examples of what mainstream animal behavior science considers to be anomalies—they seem puzzling at first because they don't appear to be optimizing individual fitness in the specific ways that scientists expect them to. But these anomalies are quickly absorbed into the optimality paradigm by mounting a search for plausible costs and benefits and caveats and mitigating factors that could explain how these behaviors are in fact optimal after all. Such an absorption prevents animal behavior science from escaping the paradigm of optimality thinking—these so-called anomalies are stripped of their power to provoke change.

Kuhn was optimistic that, in time and under the weight of enough anomalies, paradigms would break, either giving way to other paradigms, or causing scientists to give up that line of inquiry altogether. But Kuhn's optimism seems to be founded on denying the political entanglements of science. He wrote that "an apparently arbitrary element, compounded of personal and historical accident, is always a formative ingredient of the beliefs espoused by a given scientific community at a given time."[33] By emphasizing the fact that our personal experiences and historical accidents influence what we know about the world, Kuhn described something similar to the feminist science studies concept of situated knowledges (discussed in chapter 2). And yet there is a critical difference between the two: while Kuhn regards the entanglement of human perspectives and scientific paradigms as "apparently arbitrary," feminist science studies scholars understand these entanglements to be shaped by systems of power, making the political commitments of scientific paradigms the *opposite* of arbitrary.[34] Which means that escaping a paradigm is always a political endeavor, and can never be a purely scientific one. As philosopher Helen Longino described, "The feminist scientific revolution . . . will come about not because empirical anomalies accumulate and throw the current paradigm into crisis, but because changes in social values and relationships require a different way of knowing the natural world."[35]

TWO WAYS TO BREAK A PARADIGM

The paradigm of optimality thinking is inextricable from a capitalist view of the world. In a capitalist economy, class hierarchies and vast wealth inequities determine who is and isn't able to access health, stability, the freedom to shape their own lives, and the power to determine the conditions of others' lives. Capitalism justifies these economic hierarchies with the hierarchy of meritocracy, which is the fallacious notion that a person's economic conditions are determined by (or can be overcome by) nothing more than their individual capabilities and hard work, and that their positions in economic hierarchies are therefore deserved.[36] But capitalism is far from the only system of oppression that shapes our world. Indeed, the staying power of capitalism is due in large part to the fact that its hierarchies are inextricable from all manner of social and political hierarchies—of race, gender, sexuality, caste, disability, and more.[37] Hierarchy is at the heart of multiple, simultaneous, and interlocking systems of oppression, all of which depend upon each other.[38]

When applied to animal behavior science, optimality thinking depends upon the existence of a fitness hierarchy across individual animals (or groups of animals)—there can be no survival of the fittest if individuals (or groups) cannot be ranked by their fitness. Its fundamental dependence on hierarchy makes the paradigm of optimality thinking, and therefore mainstream animal behavior science itself, very amenable to naturalizing systems of oppression. When the world we live in is dominated by social structures that operate as though some lives are more valuable than others, and when scientists are unaware of how the knowledge they produce is shaped by those social structures, then the fitness hierarchies of a worldview rooted in optimality thinking end up aligning with, reflecting, and naturalizing the hierarchies of racism, ableism, sexism, homophobia, transphobia, capitalism, and more.

Standpoints that are anchored in opposition to each of these systems of oppression can thus offer us pathways out of the paradigm of optimality thinking. In the next two sections, we travel on two such pathways—toward abundance, and beyond fitness—that are grounded in Indigenous, queer, and Marxist standpoints. In the rest of this book, we broaden our perspectives to include anti-ableist and anti-racist standpoints as well.

TOWARD ABUNDANCE, FROM AN INDIGENOUS STANDPOINT

In failing to question the assumption of optimality, behavioral economists and animal behavior scientists alike take for granted that optimality is a rational response to the necessary condition of *scarcity*. When resources are scarce, it seems rational to use them efficiently. Indeed, though the word "economics" may be defined in many ways, several of these definitions describe the field as the study of scarcity.[39] But are resources always scarce? It can be hard to imagine otherwise, in part because these narratives of resource scarcity and subsequent competition are so pervasive in current, mainstream understandings of both humans and animals. Nonetheless, other views are possible. In an essay titled "The Serviceberry: An Economy of Abundance," ecologist Robin Wall Kimmerer roundly rejects the assumption of resource scarcity and offers a different perspective.[40] Kimmerer's perspective is partly situated in the social and ecological traditions of the Citizen Potawatomi Nation, a sovereign tribal nation of which she is a member, headquartered in what's now known as Oklahoma.

In this essay, Kimmerer described her experience of gathering serviceberries: "The bushes are laden with fat clusters of red, blue, and wine purple in every stage of ripeness—so many, you can pick them by the handful. I'm glad I have a pail and wonder if the birds will be able to fly with their bellies as full as mine."[41] Though she easily could have framed her experience as competing with, or losing some of the berries to, the birds, Kimmerer chose not to adopt the lens of scarcity-based capitalist economics to make sense of her experience of foraging, alongside the birds, on serviceberries. Rather, she looked through the lens of what she called an *abundance economy* or *gift economy*. Gift economies are underlain by relationships of mutuality—giving and receiving, rather than a meticulous accounting of credit and debt fueled by a tendency toward self-centered accumulation.[42] In a gift economy, as Kimmerer put it, "wealth is understood as having enough to share, and the practice for dealing with abundance is to give it away. In fact, status is determined not by how much one accumulates, but by how much one gives away. The currency in a gift economy is relationship, which is expressed as gratitude, as interdependence and the ongoing cycles of reciprocity."[43]

Scarcity thinking can provoke people into wanting or needing more—when we believe that there isn't enough for everybody, it can seem essential for each of us to keep as much as possible for ourselves, and our wants and needs grow indefinitely. In contrast, when we experience abundance, we're not as scared of running out of resources and feel less of a need to hoard. When we can trust that resources are abundant, our wants and needs can become smaller, which only increases our sense that there is plenty for all of us. While we cannot simply wish ourselves into wanting or needing less, abundance thinking suggests that our wants and needs emerge from both our material conditions and our relationships, and that when we shift from scarcity thinking to abundance thinking, we may recognize that the condition of scarcity is only rarely a "natural fact," and far more often an "economic assumption" manufactured by capitalism.[44] In her 2013 book *Braiding Sweetgrass*, Kimmerer imagined this shift in mindset by picturing what it would be like to gather groceries in a market where vendors gifted their wares rather than selling them: "Had all the things in the market merely been a very low price, I probably would have scooped up as much as I could. But when everything became a gift, I felt self-restraint. I didn't want to take too much. And I began thinking of what small presents I might bring to the vendors tomorrow."[45] In a worldview that is rooted in abundance thinking, individualism, self-interest, and optimality thinking quickly fade into irrelevance.

BEYOND FITNESS, FROM QUEER AND MARXIST STANDPOINTS

Evolutionary ecologist and trans woman Joan Roughgarden began her 2004 book *Evolution's Rainbow* by describing her very first experience of the San Francisco Pride Parade, just before she began her gender transition. She recounted first being stunned and awed by the sheer diversity in human gender expression and sexuality she encountered at the parade. Soon after, she noticed a tension: on the one hand, this tremendous diversity was fascinating, but on the other hand, scientific understandings of human and animal behavior suggested that this diversity shouldn't exist. Roughgarden wrote, "How, I wondered, does biology account for such a huge population that doesn't match the template science teaches as normal? When scientific theory says something's wrong with so many people, perhaps the theory is wrong, not the people."[46]

Mainstream animal behavior science regards evidence of queer sex (or, as scientists have taken to calling it, same-sex sexual behavior) in animals as paradoxical. A 2009 review on same-sex sexual behavior by evolutionary biologists Nathan Bailey and Marlene Zuk summarized the dominant approach of research on the subject as follows: "In the scientific literature, emphasis is often placed on the 'apparent paradox' that same-sex sexual behavior presents. It 'appears to be inconsistent with traditional evolutionary theory' and 'seems to violate a basic "law" of nature: that of procreation.'"[47] Or consider, for example, the title of a 2016 paper: "Same-Sex Sexual Behavior in Crickets: Understanding the Paradox."[48]

A valid response from queer scholars and activists to mainstream science's denouncement of queerness as paradoxical, abnormal, or unnatural is to insist that being "natural" should not have to justify queerness. It would not matter, in other words, even if there were no evidence of queer sex or nonbinary sexes among animals; people should still be entirely free and welcome to be queer, trans, intersex, and nonbinary.[49] That said, there is an abundance of material evidence demonstrating that what human societies today label "queer," "trans," "intersex," or "nonbinary" is perfectly common, even ubiquitous, across animals.[50] This abundant evidence only amplifies Roughgarden's assertion that perhaps mainstream science is what's wrong here, not the animals.

Why is same-sex sexual behavior so often described as a paradox? Because engaging in same-sex sexual behavior is thought to come at an obvious cost to reproductive output and therefore fitness, especially in a hypothetical comparison to engaging in different-sex sexual behavior. Thus, within the paradigm of optimality thinking, natural selection is expected to eliminate same-sex sexual behavior over evolutionary time *unless* the behavior provides individuals with other benefits.[51] But it is clear from the extensive evidence of same-sex sexual behavior in animals that natural selection has not eliminated it. Thus, optimality thinking tells us same-sex sexual behavior must have some fitness benefits that are unrelated to producing one's own offspring; scientists have devoted their time and resources to determining what these benefits might be.

In other words, mainstream animal behavior science is so steeped in heteronormativity that it becomes impossible to see queerness *as* queerness. In order to search for these behaviors' fitness benefits, scientists have had to get creative with their interpretations of queer behavior: mounting another individual of the same sex is interpreted as a show of dominance or aggression (as in giraffes, hopping mice, and flour beetles, among others), stimulating another individual's genitals is interpreted as play or a greeting (as in bonobos, bottlenose dolphins, vampire bats, and gray whales), and—this one's especially creative—swallowing another individual's semen is seen simply as a way of obtaining nutrition (as claimed to be true in orangutans by at least one researcher).[52] In his book *Biological Exuberance*, biologist and linguist Bruce Bagemihl described this attitude as "anything but sex," and further remarked that "in many cases, such 'explanations' are not so much genuine attempts to understand the phenomenon as they are ways of denying its existence in the first place."[53] As an especially clear example of this attitude and its effects on how animal behavior is interpreted by scientists,

Bagemihl quoted the biologist Valerius Geist reflecting, in 1975, on his earlier observations of male bighorn sheep:

> I still cringe at the memory of seeing old D-ram mounting S-ram repeatedly. . . . true to form, and incapable of absorbing this realization at once, I called these actions of the rams *aggresosexual* behavior, for to state that the males had evolved a homosexual society was emotionally beyond me. To conceive of those magnificent beasts as "queers"—Oh God! I argued for two years that, in [wild mountain] sheep, aggressive and sexual behavior could not be separated. . . . I never published that drivel and am glad of it. . . . Eventually I called the spade a spade and admitted that rams lived in essentially a homosexual society.[54]

The overt homophobia of Geist's early reactions to bighorn sheep is relatively rare among contemporary animal behavior scientists. However, dominant society, and, by extension, dominant science, remains *implicitly* homophobic enough that queer sex among animals is still seen as an anomaly in search of an explanation.

In stark contrast to mainstream science's "anything but sex" and "paradox in search of resolution" attitudes toward same-sex sexual behavior in animals, a queer feminist approach to animal sex science invites us to remain ever-curious about difference and ever-compassionate toward the "other," valuing the lives of a diversity of organisms and refusing to pass judgment on behaviors merely because they deviate from one's expectations or preferences. It therefore insists that scientists embrace and learn from the vast diversity of "oddities" and outcasts that populate the natural world instead of trying to fit them into narratives that are more acceptable to the mainstream.[55] This invitation is explicitly political; valuing difference does not mean ignoring how structures of power so often determine who or what is seen as "normal" versus "abnormal." A queer feminist approach thus empowers animal behavior scientists to empathize with the natural world in ways that arc toward justice for humans and animals who have been, and remain, oppressed. To quote environmental humanities scholar Nicole Seymour, bringing queer approaches into how scientists study nature may be understood as "a twofold project . . . to both revalue nature where it is unjustly devalued, and challenge 'nature' where it becomes the rationalization for injustice."[56] In keeping with this second charge: a queer feminist approach to understanding same-sex sexual behavior helps us break out of the paradigm of optimality thinking.

In a 2019 paper titled "An Alternative Hypothesis for the Evolution of Same-Sex Sexual Behaviour in Animals," ecologist Julia Monk and colleagues (including Max Lambert, from chapter 1, and Ambika) outlined how mainstream science's understanding of same-sex sexual behavior as a paradox depends upon two assumptions that are clearly aligned with heteronormativity, if not homophobia. First is the assumption that same-sex sexual behavior is *necessarily costly* in terms of lost opportunities for reproduction. However, this assumption is rarely tested. Monk and colleagues put it thus: "Until now, evolutionary biologists have asked why same-sex sexual behavior has evolved and how it can persist despite the obvious costs. We counter by questioning whether the obvious costs to same-sex sexual behavior are really so obvious after all."[57]

By treating same-sex sexual behavior as obviously costly and therefore paradoxical, mainstream scientists seemed to disregard the possibility that individual animals can and do engage in *both* same- *and* different-sex sexual behavior—a particular erasure of bisexuality or pansexuality, if you will.[58] For example, in several species of insects, including water striders, wasps, and beetles, individuals who mate more often with same-sex partners also, on average, tend to mate more often with different-sex partners.[59] And in what is likely the most comprehensive study of same-sex sexual behavior in the wild to date, ecologist Jackson Clive and colleagues showed that male-male sexual behavior in Rhesus macaque monkeys is not only more commonly observed than male-female sexual behavior, but also that engaging in male-male sexual behavior does not negatively impact these males' offspring production.[60] In general, there is little systematic evidence to suggest that engaging in same-sex sexual behavior necessarily reduces an individual's fitness.[61] Thus, even assuming that fitness is all that matters, as the paradigm of optimality thinking insists one does, the argument that same-sex sexual behavior is obviously paradoxical holds no water. Monk and colleagues ultimately proposed that same-sex sexual behavior persists in many animals not because its vast costs are outweighed by even bigger benefits but because, in many ecological and social contexts, it simply isn't very costly at all.

The assumption that same-sex sexual behavior is substantially and obviously costly belies a homophobic notion of queer sex as somehow excessive or wasteful. Moreover, this assumption exists alongside a second, mirrored assumption, also rarely stated explicitly: that different-sex sexual behavior is *not* excessive, or that mating with a different-sexed individual results in the *efficient* production of offspring. In fact, empirical evidence suggests that animals engage in different-sex sexual behavior far more often than is necessary for fertilization. "Heterosexual pairs of goshawks and American kestrels mate 500–700 times for each clutch of eggs they produce," for instance.[62] (At the time of writing, Ambika has been watching several Cooper's hawks outside her kitchen window for a few minutes each day; even in this brief period of observation, she can confirm that these hawks have been copulating often and "inefficiently." Maybe some animals simply enjoy sex?)

Thus, there is a deeper assumption inherent to optimality thinking that a queer feminist perspective insists we interrogate: the assumption that animals necessarily behave in ways that optimize their reproductive success. What if maximizing individual reproductive success is not all that matters to an animal's life and a population's evolutionary trajectory? For us to take this possibility seriously, we have to return to the idea that we considered when delving into the definition of natural selection: that instead of selecting for the "fittest" individuals, natural selection simply selects for "fit enough" individuals.[63] In other words, it does not matter if an animal produces *the most* offspring; all that matters for the persistence of populations and individual animals' own genetic legacies is that they produce *some* offspring. This perspective makes ample room for animals to engage in all kinds of behaviors that do not optimize fitness, including lots of or little sex with individuals of any and all sexes. When we let go of optimality thinking, queer sex in animals is no longer a paradox.

But queer feminist perspectives push us even further beyond rejecting "survival of the fittest" in favor of "survival of the fit enough"—they also push us to question the primacy we accord to reproduction and genetic legacy in our current conceptions of evolution by natural selection.[64] Mainstream science offers us precious few frameworks for thinking about the evolutionary impact of an individual's behavior in terms unrelated to the production of offspring or to leaving a genetic legacy. Queer feminist perspectives insist that we think more expansively.

Whenever Melina presented her and Max's work on queer frog sex to biologists (see chapter 1), she was surprised by some of the strong pushback she received, even from biologists who self-identified as social justice advocates. It wasn't until working on this book that Melina realized what was preventing otherwise open-minded and highly intelligent scientists from embracing queer feminist approaches to frog sex. Biologists objected to her and Max's analyses largely because of the unnamed assumption inherent to the paradigm of fitness optimization: that reproduction is *the singular purpose* of animal life. If one assumes reproduction is the singular purpose of an animal's (or every animal's) life, and if a chemical pollutant is interfering with frog reproduction, for example, then it logically follows that said chemical *must* be causing species decline. It was at this point that Melina realized how deeply heterosexist mainstream biology is—a queer feminist approach to animal behavior would never assume that reproduction is the central purpose of life.

The existence of queer chosen family (introduced in chapter 3) demonstrates how a vast number of social interactions and relationships, beyond reproductive ones, can be central to a human or an animal's life. Thus, a queer standpoint, and the concept of queer chosen family in particular, helps us see beyond fitness altogether. It is undeniable that organisms living in a community (that is, all organisms) substantially affect one another through their mutually influential interactions within and across species, just as queer chosen family do. Moreover, these spacious understandings of relationships offered by queer standpoints interlock with the Marxist feminist concept of social reproduction, which together empower us to fully set aside mainstream science's assumption that the production of offspring (and, by extension, reproductive sex) is the only thing that organisms live for.

In their trenchant critiques of the economic system of capitalism, Marxists have tended to focus on the worker, without whom capitalism could not function. Capitalism needs to maintain a steady supply of workers who must sell their labor in exchange for wages (and needs to pay them little enough that workers have no choice but to continue working and spending their wages on basic needs). What Marxist feminists have rightfully added to this analysis is that the worker does not simply materialize out of thin air at the factory door. As historian Tithi Bhattacharya wrote:

> If workers' labor produces all the wealth in society, who then produces the worker? Put another way: What kinds of processes enable the worker to arrive at the doors of her place of work every day so that she can produce the wealth of society? What role did breakfast play in her work-readiness? What about a good night's sleep? We get into even murkier waters if we extend the questions to include processes lying outside this worker's houschold. Does the education she received at school also not "produce" her, in that it makes her employable? What about the public transportation system that helped bring her to work, or the public parks and libraries that provide recreation so that she can be regenerated, again, to be able to come to work?[65]

All these other things that produce the worker are what Marxist feminists call "social reproduction." By highlighting the essential role of many different kinds of labor in maintaining society and rejecting the idea that women perform this labor "naturally," the concept of social reproduction helps us expand our sense of what's required to sustain human life.

Applying the concept of social reproduction in the realm of animal behavior science can greatly expand what we think of as important in an animal's life too, allowing us to recognize that having sex to produce offspring need not be the singular purpose of an animal's existence. Because it's not just the work of producing offspring that matters to an animal population's persistence. An individual's impact on their population also depends on their social and ecological relationships with other individuals in the same or different species. A female moth chooses where to lay her eggs by copying the choice of another female, and thereby allows for the other female's choice to influence her own offspring's lives. Songbird chicks learn how to sing by listening to the adults that tend to their nest, who may or may not be their biological parents. A beaver who builds a dam across a river changes the lives of the fish in that river, and the insects that the fish feed on, and the birds that feed on the fish. Thanks to the capitalist underpinnings of optimality thinking, mainstream animal behavior science insists that these diverse impacts that organisms have on one another are only important insofar as they contribute to optimizing an individual's fitness. By shedding the narrow constraints of what capitalism deems important and instead situating ourselves alongside a diversity of marginalized standpoints, we can start to see that all these impacts are a part of these individuals' evolutionary legacies regardless of their relationships to fitness, simply because they shape the conditions of their own and others' lives.

STRONG OBJECTIVITY AND CO-CONSTITUTION

The more people there are doing science from explicitly different perspectives, the richer our scientific understanding of the world becomes. After all, the more pieces of partial knowledge we have, the closer we may come *collectively* to a sense of what the world might be. But in order to accept situated knowledges fully—and specifically, to appreciate the capacity of marginalized standpoints to help break out of paradigms—scientists have to give up the idea that mainstream science is a uniquely neutral way of determining the objective truth about the world, and must instead learn how to hold multiple truths simultaneously. Additional concepts from feminist science studies—strong objectivity and co-constitution—offer alternative ways to proceed.

At first, losing the claim to neutrality and objectivity seems like a huge blow to what science can achieve. The idea that one cannot know the objective truth about the world seems entirely foreign to the work that scientists do. Feminist science studies scholar Sandra Harding introduced the concept of "strong objectivity" in 1995 as a way to *embrace* the unavoidable and inevitable partiality of all perspectives.[66] Strong objectivity is the idea that scientists' efforts at objectivity are actually strengthened when they explicitly acknowledge their particular social locations, and overtly attempt to see things from different points of view. Strong objectivity implies that scientists will actually produce closer-to-complete understandings of the world when they admit that, just like those of any other human, their perspectives are necessarily limited and historically situated and unavoidably affect how they pursue research and interpret results. Scientists who embrace strong objectivity will thus elect to engage with multiple perspectives regarding the same phenomenon. More recently, Black feminist geographer Katherine McKittrick has cautioned mainstream scientists against superficially engaging marginalized perspectives without adequately accounting for the specific histories behind, and violent consequences of, who and what has been marginalized. Instead, McKittrick emphasized that collaborations between different practitioners and different ways of knowing are most fruitful when they are intentionally nonhierarchical.[67]

One way to understand strong objectivity is to imagine the truth as a multidimensional blob.[68] For simplicity's sake, in figure 4.2, truth is pictured as a cylinder.[69] We can describe and come to know this truth blob *only* by shining light on it from different angles and positions and looking at the

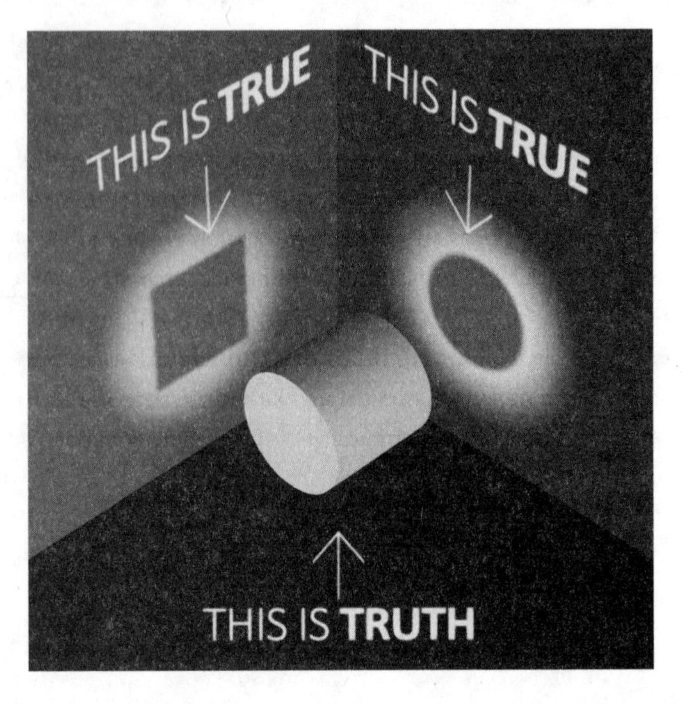

Figure 4.2
Light shining from two directions onto a single three-dimensional cylinder, yet casting different shadows onto two perpendicular planes: a visual metaphor for several important feminist science studies concepts.

shadows cast on different planes—our knowledge is therefore necessarily partial *and* requires an observational apparatus, in this case a light and a surface on which the shadow is cast. Moreover, the shape of the shadow we see depends on where we are casting light *from*—all knowledge is situated (see chapter 2), and we don't entirely get to choose the position from which we see the blob. Standpoint theory (also introduced in chapter 2) reminds us that, when we're looking at the truth blob, some locations from which one might look are available only to those of us whose political consciousness has developed in response to being marginalized. And finally, strong objectivity means that our best chance of understanding the truth blob comes from shining light on it from as many different locations as possible, communicating what we each see, and then considering all our perspectives together while paying attention to, and accounting for, how histories of marginalization may lead some perspectives to seem blurrier or less bright than others.

These feminist science studies concepts of situated knowledges, standpoint theory, and strong objectivity do not imply that all people's truths are equally acceptable, however, because feminist science studies remains attentive to power. Those who claim that injecting disinfectant into one's veins will cure COVID-19, for instance, have a particular political stake in believing that to be true, and indeed may benefit from the mismanagement of the COVID-19 pandemic. Meanwhile, marginalized people who disproportionately bear the pandemic's burdens have good reason to be skeptical of the claim that injecting disinfectant into one's veins will cure COVID-19, and to pay attention to the harmful material consequences of such "treatment."[70] The truth blob can never be divorced from its context, in other words, so truths that benefit the powerful can only be revealed as such. By remaining alive to, and working to understand, the political stakes of any and all claims to scientific truth, scientists have a far better chance of aligning with justice and against harm.

While strong objectivity helps us conceive of a way of doing science that makes room for multiple perspectives, it nonetheless still suggests that there *is* something like the truth out there that we can potentially uncover together, even if none of us can uncover it individually. The feminist science studies concept of *co-constitution* unsettles this notion of a truth that is separate from us, waiting to be revealed. Instead, the concept of co-constitution posits that the very act of observing a phenomenon is what generates the phenomenon.[71] Phenomena do not preexist the observer and their observational apparatuses—the truth blob doesn't even exist before we shine a light upon it, and the truth blob that comes into being as we shine a light doesn't just look different but *is* different, based on our particular light-producing and shadow-capturing tools. The observer and the observed are *always already becoming, together*—they are, in other words, always and continuously *co-constituted.* Co-constitution, or the inseparability of any facet of nature from what and how we come to *know* about that facet of nature, is foundational to feminist science.

Indigenous feminist science studies scholar Kimberly TallBear has noted how these contemporary concepts of co-constitution sound to her an awful lot like the Lakota/Dakota cosmologies within which she was raised:

> [The] looping and not linear account that we are all of us—humans and nonhumans—a networked set of social-biological relations . . . resonates with what [Dakota scholar] Vine Deloria, Jr. called an "American Indian metaphysic." . . . In his 2001 essay "American Indian Metaphysics," Deloria wrote that the best description of that term is: "the realization that the world, and all its possible experiences, constituted a social reality, a fabric of life in which everything had the possibility of intimate knowing relationships because, ultimately everything was related." Is it too easy a comparison to say that Western thinkers are finally getting on board with something that is closer to an American Indian metaphysic?[72]

While demanding recognition for, in this case, Dakota standpoints, TallBear did not insist that only Indigenous people are entitled to use the concept of co-constitution, even though they may have extensive histories of doing so. Instead, TallBear saw an opportunity for a "greater scope for bringing indigenous voices to the conversational table."[73]

Beautifully, this table at which we're talking about co-constitution as foundational to our worldview is a crowded one. We may each come to recognize the importance of co-constitution through different routes. Some of us may have found value in the work of past scientists who have pushed against the mainstream, while others of us may have followed a path that had little to do with science directly. Some of us may arrive here via a largely Marxist analysis, others of us may adopt a primarily feminist perspective. Some of us may find ourselves believing in co-constitution thanks to a Buddhist understanding of interbeing, and yet others of us might be rooted in Indigenous cosmologies. But now that we are all here, it's our job to learn from one another, in a way that rejects hierarchy. As we find what unites us and share our complementary understandings of the world, we can begin to dismantle dominant paradigms together. Collectively, our diverse perspectives lead us to a different, feminist paradigm for animal behavior science—one that is grounded in co-constitution. We continue building this different paradigm throughout the rest of *Feminism in the Wild.*

5 UNDOING ADAPTATIONISM

LEAPING LIZARDS!

Evolutionary biologist Jonathan Losos thinks of himself as an adaptationist. His approach to studying nature, like that of so many scientists studying the evolution of animal behavior, is to pay close attention to what animals look like and how they behave, and then ask: "How might this animal's morphology and behavior be a result of adaptation by natural selection?" Adaptations, you'll remember from chapter 1, are the uncanny alignments between organisms and their environments. The theory of natural selection was developed to supply an explanation for adaptations—a close match between an organism and their environment can arise when generations of natural selection increase (or even optimize) the fit between the two.

In his decades of research on the evolution of anole lizards, Losos has found plenty of evidence for the story that these lizards' morphologies are near-perfectly matched to their particular environments.[1] Some species of anole lizards live on the trunks of trees and near the ground, while others live much higher, amid the branches of tree canopies. Having measured thousands of anole lizards across the Caribbean, Losos consistently found that species that live on broad surfaces closer to the ground have longer limbs, whereas species that live on narrower surfaces farther from the ground have shorter limbs (among many other differences). To explain this pattern of variation, Losos has long told an adaptationist story in which this close match between limb lengths and running surface was a result of natural selection optimizing the length of these various lizard species' limbs for maximal running performance on each type of surface.

As a PhD student in the late 1980s, Losos collected more experimental data in support of this adaptationist hypothesis. He prompted lizards with different limb lengths to run on a homemade lizard racetrack, to measure the speeds at which they ran. He found that longer-legged lizard species ran quicker on flat, broad surfaces, but often lost their footing on narrower structures. Shorter-legged lizard species, in contrast, could maneuver more easily on narrow surfaces. This variation among lizard species in their performance on different surfaces in the lab matched their choices of habitat in the wild—long-legged lizard species lived on broad surfaces (tree trunks and the ground) where they could run more quickly, and short-legged lizard species lived on narrower surfaces (branches) where they could maneuver more easily. Based on this match, Losos felt confident drawing a straightforward adaptationist conclusion: natural selection had optimized running performance on different surfaces by optimizing limb length, and different anole lizard species were thus adapted to living in different parts of trees.

Amid this adaptationism, a single lizard sowed a seed of doubt in Losos's mind. "I caught a lizard that was missing most of his fore leg," he described. "And just as a lark, really, I put him on the racetrack anyway, and was surprised how fast he ran. He definitely didn't run as fast as other lizards, but he did alright."[2] This lizard unsettled Losos's sense of how natural selection had determined these lizards' morphology. He had assumed that an individual lizard's capacity to survive depended upon their performance, which in turn depended upon their morphology, and that even tiny differences in morphology between lizards had consequences for the differences in their fitness. But this three-legged lizard upended Losos's assumption. "Here I am talking about what a millimeter's difference in leg length makes [to sprinting and maneuvering in different species of lizards], and here's a lizard missing most of its leg, and it can still run pretty well!" Losos said. "We think performance is so important, and [yet] there's some three-legged lizards out there doing just fine."

In the decades that followed, Losos continued to observe all kinds of lizards missing parts of their limbs or even whole limbs, many of whom continued to surprise him. "There was a lizard on our study islands in the Bahamas that I was trying to catch, and she was being very difficult. She was very nimble, and it wasn't until I had her in my hand that I realized her entire hind leg was missing! I hadn't even noticed as I was chasing her, she was so adept," Losos told us. Because this lizard was part of a long-term study and had been measured a year ago, Losos knew that she used to have all four limbs. Losing *an entire leg* had not imposed an obvious cost on this animal's survival (figure 5.1).

In the world of optimality thinking, where lizards with all their limbs intact are seen as optimal solutions to the problem of existing in their particular environment, it seems obvious that a lizard who loses a limb will suffer as a result of the loss—he would no longer be optimal. But in a world in which optimality is irrelevant, when a lizard loses a leg, she has options. She may change the precise patterns by which her remaining limbs and her tail move, allowing her to run quickly enough again. She may move into an environment that better supports these new patterns of movement, or she may move in search of prey that are easier to catch or into a spot more sheltered from predators. Her life changes, for sure, but she is by no means necessarily relegated to a lesser life.

Losos is still an adaptationist (though, as he said, "not a hardcore, irredeemable one"). So when he opens up his datasheet of seventy-five (and counting) haphazardly gathered observations of wild lizards with missing limbs that nonetheless seem to be thriving, he experiences some cognitive dissonance.[3] Losos now regards these lizards as "a cautionary tale for people who, like me, have assumed that performance is critical."[4] He realizes he cannot simply assume that losing a limb necessarily reduces an animal's fitness, or that one can neatly infer variation in fitness from variation in what animals look like or how they behave. And these cracks in Losos's commitment to optimality thinking stem from recognizing that "organisms are not automatons." Instead, animals "are very flexible [and] self-regulating . . . that somehow a lizard with three legs changes what it does to be able to survive."[5]

With Kuhn's framework of paradigms in hand (see chapter 4), we can recognize Losos's cognitive dissonance as the moment at which a paradigm is confronted with apparently anomalous data that, at first, don't seem to make sense within the prevailing way of understanding the world. If lizard

Figure 5.1
An X-ray image of a wild brown anole lizard missing a hindlimb, from the island of Abaco in the Bahamas. Image: Jonathan Losos.

limb lengths are adapted to specific running surfaces—that is, if they are optimal solutions to the problem of living in different parts of trees—then how can losing an entire limb not have severe fitness consequences? The paradigm in question here is not just optimality thinking but also its conjoined twin, adaptationism—while optimality thinking insists that adaptations are a result of natural selection working to optimize fitness, adaptationism is the tendency to see all biological traits as adaptations in the first place.

We also know, from feminist critiques of Kuhn's framework, that if we want to break out of the dual paradigm of adaptationism and optimality thinking, the odds are stacked against us. Without deliberate intervention, a paradigm will simply absorb any anomalies into itself. Here, such an absorption could look something like the argument that lizards without a hindlimb are in fact just doing all they can to make the best of a bad situation, surviving as long as they can and salvaging what they can of their reproductive success in order to optimize their fitness anew, given their changed circumstances. Moreover, adaptationism would insist that the lizards' changed behaviors in response to limb loss are themselves somehow an optimal result of adaptation by natural selection. As we saw in chapter 4, this logic cannot be disproven; breaking free from it requires questioning the assumption of optimality itself, as well as the assumption that all behaviors are adaptations. Last, and crucially, feminist science studies tells us that a paradigm-breaking intervention must be rooted in an interrogation of the political stakes of the scientific concept at hand—empirical arguments alone cannot release us from a paradigm.

Insights from the realm of disability justice clarify the stakes of the dual paradigm of optimality thinking and adaptationism. An anti-ableist standpoint questions the centrality of fitness hierarchies in optimality thinking by reminding us—as the thriving three-legged lizards did—that a trait that appears to be a disability may not in fact compromise fitness at all.[6] When animal behavior scientists assume that disabled or differently abled animals are necessarily unfit or less fit, they shut themselves off from exciting research findings and greater biological understandings. As queer feminist disability theorist Alison Kafer put it: "To eliminate disability is to eliminate the possibility of discovering alternative ways of being in the world."[7] Disability doesn't have to be a weakness that will or ought to be selected against.[8]

That said, an anti-ableist refusal of fitness hierarchies does not require that one sees nature as some fantastical utopia with no suffering or death. Rather, to reject fitness hierarchy as an organizing principle for animal behavior science is to think carefully about how suffering and death are distributed across humans and all organisms. It means accepting disability, insisting that it is not inherently *less than* while also refusing to romanticize it, particularly when disability is imposed by external, more powerful forces or comes with a considerable amount of pain or requires substantial care work.[9] As critical disability studies scholar Sunaura Taylor said of her congenital disability caused by in utero exposure to military toxicants: "I hate the military and love my body."[10] An anti-ableist analysis accepts that death and suffering are a part of life while highlighting the uneven distributions of suffering and death generated by systems of power.[11] Crucially, anti-ableist approaches refuse to justify or naturalize these uneven distributions by uncritically mapping them onto fitness hierarchies.

Which is to say, an anti-ableist perspective reveals that one cannot break out of optimality thinking by simply gathering data to show that varied individuals, some seemingly able-bodied and others seemingly disabled, are all equally fit. In fact, the logic of such an approach would be entirely consistent with the paradigm of optimality thinking—if all variation is already optimal, then all variants *must* be equally fit, and the job of a scientist operating within the paradigm is to come up with the right combination of variables to measure that would demonstrate equal fitness. Such a search could continue ceaselessly, but to search interminably for data that might eventually demonstrate equality while clinging to a worldview that fundamentally depends upon hierarchy is not the same as rejecting hierarchical thinking altogether. To break free from naturalizing unjust hierarchies, one must altogether refuse to transform biological variation into fitness hierarchies—without hierarchies to map onto, variation is just variation.

It's worth remembering that animal behavior scientists are invested in fitness hierarchies and fitness optimization for an explicit, scientific reason:

to explain the existence of adaptations, or the near-perfect fits between organism and environment. Thus, without further intervention, even if mainstream animal behavior science were to divest from hierarchies and optimization, the field would remain invested in the adaptationist tendency to assume that biological traits represent an *as-close-to-perfect-as-possible* fit between organism and environment. It would therefore also remain entangled with ableism, which disability studies scholar Fiona Kumari Campbell has defined as "a network of beliefs, processes, and practices that produces a particular kind of self and body . . . *perfect*, species-typical and therefore essential and fully human."[12] Undoing the ableism embedded into mainstream animal behavior science therefore requires confronting the paradigm of adaptationism directly.

IF NOT ADAPTATIONISM, THEN WHAT?

The idea that biological traits are adaptations in fact *preceded* the theory of natural selection. Centuries before Darwin, Western naturalists and philosophers had noticed and pondered what they saw as the widespread pattern of near-perfect matches between organisms and their environments. Indeed, organisms appeared to these "natural theologians" as having been *designed* to persist in the environments they occupied, and only divine forces were thought to be powerful enough to produce these patterns of adaptation—this is known as the "argument from design." Historian Jessica Riskin described the 1600s perspective of natural theology as follows: "If God was an engineer, his machinery must be perfectly designed: means ideally adapted to the ends, structures flawlessly suited to the functions."[13]

The theory of natural selection wrested this capacity for creating adaptations from the hands of God. As evolutionary biologist Ernst Mayr put it, "to have been able to provide a scientific explanation of adaptation was perhaps the greatest triumph of the Darwinian theory of natural selection. After 1859 it was no longer necessary to invoke design, a supernatural agency, to explain the adaptation of organisms to their environment."[14] Given this history of divine explanations for adaptation, it can perhaps feel necessary to think of natural selection as a relentlessly optimizing force (or invisible hand) that produces near-perfect adaptations—otherwise, it's hard to outdo an all-powerful god.

While it's true that the theory of natural selection was developed to explain adaptations and in doing so, provided a material cause for phenomena that were once seen as divinely created, there is nothing about the process of natural selection itself that necessarily connects it to adaptation.[15] The process of natural selection requires only that some individuals survive and reproduce more than others. It does not require that this differential survival and reproduction is always, or even often, either *due to* adaptation or *leads to* adaptation. Natural selection remains the best material explanation scientists have for how biological traits become adaptations. But this fact need not compel animal behavior scientists to assume that any and every biological trait is, first and foremost, an adaptation.

Though lip service has long been paid in mainstream science to the possible existence of nonadaptive explanations for how traits evolve, such explanations are rarely taken seriously. As Marxist evolutionary biologists Stephen Jay Gould and Richard Lewontin put it, in a scathing 1979 critique of adaptationism, "the admission of alternatives in principle does not imply their serious consideration in daily practice."[16] A rebuttal to Gould and Lewontin from Mayr only further proved their point. Mayr's 1983 paper titled "How to Carry Out the Adaptationist Program?" included the following prescription for how to study biological traits: "when one [adaptationist] explanation of a feature has been discredited, the evolutionist must test other possible adaptationist solutions before he [*sic*] can resign and say: This phenomenon must be a product of chance."[17] Not only did Mayr consign any evolutionary process that isn't adaptation by natural selection to "chance," but he also described no end to the scientist's search for adaptationist solutions, exactly as Gould and Lewontin had suggested. Indeed, the end remains elusive—in 2015, philosopher Elisabeth Lloyd continued to wonder whether the adaptationist approach "*in practice* allows nonadaptive explanations ever to win the day."[18]

Steeped in an adaptationist worldview, it can be difficult for biologists to come up with nonadaptive hypotheses even when they're trying to do exactly that. To illustrate, let's consider how the *Animal Behavior* textbook discusses the clitoris of the spotted hyena. Spotted hyena clitorises are about eight inches long and can become erect, leading many biologists to refer to them as "pseudopenises." Spotted hyenas not only urinate, copulate, and give birth through these organs but also closely inspect one another's erect clitorises during social interactions. Compared to those of other mammals, spotted hyena clitorises are unusual in both what they look like and how they are used.[19] *Animal Behavior* opens a section on spotted hyena clitorises, titled "The Adaptive Function of a Strange Display," thus:

> Before we examine the many potential adaptive explanations for the pseudopenis display, we must consider the possibility that the pseudopenis is not adaptive per se but instead developed as a by-product of a change that had other kinds of positive fitness effects, and that may have also resulted in high testosterone levels in females (by-product hypothesis). Perhaps the pseudopenis is simply a developmental side effect of the hormones present in newborn spotted hyenas that are favored by natural selection because they promote extreme aggression between siblings, enabling them to compete for control of their mother's parental care? . . . Alternatively, a [heritable] change of some sort that resulted in an enlarged clitoris might have spread because it helped make adult females larger and more aggressive, but not because it helped them acquire a pseudopenis.[20]

The first half of the first sentence of this paragraph holds the promise of a *non*adaptive explanation for the unusualness of spotted hyena clitorises. However, this promise unravels quickly. You'll notice that the alternative explanations the textbook considers are in fact still adaptive, because they still explain the trait (enlarged, erectable clitorises) as arising due to fitness optimization (on testosterone levels), albeit indirectly (via selection favoring heightened aggression). But the textbook authors are unsatisfied with even these indirect adaptive explanations, as they write: "There are clearly many problems with the by-product hypothesis. For one thing, if the pseudopenis is strictly a liability, why hasn't selection favored mutant alleles that happened to reduce the negative clitoral effects that androgens . . . have on female embryos? So let's consider the possibility that the pseudopenis might have adaptive value in and of itself, despite the apparent costs associated with its development or with having to give birth through it."[21] Though the authors declare that there are obviously many problems with their proposed indirect adaptive hypotheses, they name only one. Moreover, the one problem they name is not so much a problem as an assumption combined with wishful thinking. Yes, the particular morphology of hyena clitorises have consequences for these animals (copulation can get awkward, and many first-born hyena pups die during birth), and there is some evidence from laboratory manipulations that having a smaller clitoris can mitigate these challenges, but do the textbook authors know that a hyena's clitoris is really a strict liability?[22] Do they know if anyone has measured the fitness consequences of variation in clitoris size in wild hyena populations?[23] Did they consider whether the survival of a first-born pup affects hyena parents in other ways that might lower their overall fitness? Maybe the energy expended in rearing that first pup would substantially increase the parents' chances of dying before they reproduced again. In short, we do not know whether "the pseudopenis is strictly a liability." And even if it were, that in and of itself does not imply that genetic mutations must exist that somehow magically remove all the costs of an enlarged clitoris, while preserving all its benefits—that remains wishful thinking.

It might feel satisfying, when critiquing adaptationism, to poke these obvious holes into specific adaptationist arguments. But in doing so, we have in fact fallen into the trap of the paradigm. By disputing a specific adaptationist accounting of the costs and benefits of hyena clitorises, we have ended up arguing *within* the paradigm, on adaptationist terms. We have tacitly accepted that the trait is an adaptation and that accounting for its costs and benefits is the most meaningful way to understand its evolution. To break out of adaptationism, therefore, we have to consider what *Animal Behavior* entirely fails to consider: that the question of whether a spotted hyena clitoris is or isn't an adaptation *may not be important to understanding its evolution.*

This is not to say that it is *never* useful to think of traits as adaptations. Even the Marxist biologists quoted earlier agree that it can be particularly compelling to regard traits as adaptations when extremely similar traits evolve in different organisms that occupy similar environmental circumstances (a phenomenon known as "evolutionary convergence"). "It is surely no accident," Levins and Lewontin wrote, "that fish have fins; that aquatic mammals have altered their appendages to form fin-like flippers; that ducks, geese, and seabirds have webbed feet; that penguins have paddle-like wings; and even that sea snakes, lacking fins, are flattened in cross section."[24] It really does seem like there is *something* that repeatedly brings about the evolution of broad, flat, movable surfaces that animals use to propel themselves through water. The theory of adaptation by natural selection remains science's best explanation for what that *something* is. (Losos's anole lizard limbs are also an example of evolutionary convergence, suggesting that they are shaped by adaptation by natural selection to at least some degree, even if they aren't optimal).

These examples of convergence suggest that there is also *something* about these different animals' environments that remains constant across different contexts—*something* about moving through water (or on broad tree trunks or narrow branches) that remains constant no matter which organism is doing the moving. But when scientists see adaptations *everywhere* in nature, rather than just in the specific circumstances of evolutionary convergence, they are tacitly thinking of environments *in general* as static and separable, either unchanging or only changing independently of the organism.[25] And when scientists think of environments as static or separable from the organism, then it becomes easy to think of environments as posing "problems" for organisms to "solve" via natural selection. Adaptations, in this worldview, represent organisms' successful, optimal "solutions" to these problems posed by the static and separable environment.

What this adaptationist worldview ignores is how organisms' own actions—of moving through, selecting, and modifying their environments through their interactions—ensure that their environments are not static or separable from the organisms themselves. These changes to the environment, in turn, change an organism's experience of and response to the environment, which then changes their subsequent actions. And equally, one can flip this description around and examine it from the perspective of the organism's environment, which is not a monolith but rather an entanglement of forces and beings, nonliving and living, each of which or whom is altered by and alters everything and everyone with which and whom they interact. Through the narrowly constrained lens of adaptationism, this entanglement might seem like a process of optimization that fits an organism to an environment. But setting the lens of adaptationism aside, this entanglement can also be seen as an always-ongoing process of organism and environment *becoming, together*. As Levins and Lewontin put it:

> The factual difficulty of formulating evolution as a process of adaptation to preexistent problems [posed by the environment] is that the organism and the environment are not actually separately determined. The environment is not a structure imposed on living beings from the outside but is in fact a creation of those beings. The environment is not an autonomous process but a reflection of the biology of the species. Just as there is no organism without an environment, so there is no environment without an organism.[26]

Or, as feminist science studies teaches us, organisms and environments are always already *co-constituted*.

CATERPILLARS, CO-CONSTITUTION, AND CONTINGENCY

One of the joys of studying animals in the wild is that you get to watch not just the animal you intend to study but also a plethora of animals you may not even have known existed before you happened upon them. In the summer of 2015, while Ambika spent day after day in a single urban park in central Florida watching anole lizards, she couldn't help but notice all the other animals that called this park home: garter snakes, box turtles, black-and-white warblers, molting cicadas, armadillos, and stray cats, to name a few.

But more than any of these fine creatures, it was tent caterpillars who captured Ambika's attention (when she was not watching lizards, of course; figure 5.2).[27] Tent caterpillars are, admittedly, hard to ignore. They aggregate in large numbers, forming a slow-moving mass that can strip a cherry tree of its leaves in a matter of days. Their white silken tents catch the sunlight, beacons amid the branches. Ambika could have spent days on end watching these mysterious beasts, but there was work to be done gathering data on lizards, so she made a note to herself to return to watching tent caterpillars at some future time. The chance to do so arose in 2019. This time, she watched them with purpose. Her goal was to observe the animals long enough to

Figure 5.2
Eastern tent caterpillars on a *Prunus* tree at the La Chua Trail in Gainesville, Florida. Photo: Ambika Kamath.

be able to understand what the right question to ask about them was, and, given her training in behavioral ecology and evolutionary biology, she took it for granted that the right question to ask would be an adaptationist one.

But the more Ambika learned about tent caterpillars, the more unsettled she became—she simply couldn't manage to see these animals' behaviors as optimal solutions to an environmental problem, not least because their environments changed so much and because the caterpillars themselves were *causing* a lot of this environmental change! For example, if the caterpillars managed to eat many of the leaves that surround them, they would experience more direct sunlight and warmer temperatures and might have higher metabolic rates, which might in turn increase their need for food while also requiring that they travel longer distances in search of more leaves to feed on. Of course, by eating the leaves around them, the caterpillars were also changing the lives of the trees they lived on, affecting how much the plants could photosynthesize, how much energy they could store, how many fruit they could produce, and therefore how much of a source of abundance they might be for many other animals in the ecosystem who also depend on these trees.

Beyond changing their environments by eating the leaves around them, tent caterpillars also constructed their own environments more directly by, as their name suggests, building tents. Soon after hatching, groups of about two or three hundred caterpillars from an egg case would build a silken tent within which they all lived. Huddling together within these tents, the caterpillars could better buffer themselves against fluctuations in the weather. Individual caterpillars were thus not only constructing one another's environments by building a tent together, but also remained *part of* one another's environments—their evolutionary fates were and are inextricably intertwined. All of these actions, interactions, and their consequences created the conditions of tent caterpillars' lives in this generation, which themselves were the result of the conditions of previous generations, and will be the starting point of generations to come. It seemed to Ambika a strange choice to isolate a single caterpillar from this whole complex system, and ask questions about how this individual's behaviors are adaptations, or how an individual caterpillar's behaviors maximize their individual fitness. In fact, it became impossible to think of these animals' evolution *without* thinking about how organisms and environments are co-constituted.

Asking questions about animal behavior that are grounded in co-constitution rather than adaptationism is undeniably harder. One way to begin making this transition is to shift away from asking *why* questions about a biological trait and move toward asking *how* and *what* questions instead. *How* did a trait come to be this way, and *what* might it lead to in the future? What was the particular history of this population of organisms, within individual lifetimes and across generations, that led this trait to its current state? And now that the trait *is* this way, what future evolutionary possibilities become available to the organism and the ecosystem because it is so? A feminist approach to science would remind us that the evolutionary histories of traits, and indeed the descriptions of the traits themselves, will look different from different standpoints.

Nonadaptationist answers to these questions would interweave all kinds of biological processes with natural selection—fluctuations in gene frequencies across generations, the growth and development of individual animals over their lifetimes, environmental shifts and interactions, and changes in social dynamics, to name a few. A nonadaptationist approach would regard each of these processes as inextricable, rejecting the assumption that fitness optimization ultimately dominates over other processes to produce traits that are adaptations. Instead, the most important consequences of each of these processes would be their effects on one another, on the animal's whole life and on future generations, and on the entire ecosystem, *not simply* their impacts on fitness.

Ultimately, grounding one's inquiries in co-constitution rather than adaptationism makes clear that, as disciplines, animal behavior science and evolutionary biology are much closer to history than they are to economics. In his 2005 book *Logics of History*, historian William Sewell Jr. described the need for "a serious infusion of historical habits of mind."[28] Though he was speaking to social scientists, his words are just as applicable to biologists. Central to these historical habits, according to Sewell, is a theory of time, or *temporality*. Events unfold through time and actions have consequences. An action, once taken, cannot be untaken; its consequences can, however, be modified by subsequent actions. Thus, this unfolding of actions is *contingent*, which means that "every act is a part of a *sequence* of actions, and . . . its effects are profoundly dependent upon its place in the sequence."[29] As a result, "historical happenings are extremely unpredictable," and, one might add, nothing is determined or inevitable.[30] Though this degree of unpredictability may be anathema to many scientists, those who study animal behavior and evolutionary biology simply cannot escape it because their inquiries are, at their core, historical.

The contingent unfolding of events through time means *everything* to a tent caterpillar. Tent caterpillars' egg cases, each housing several hundred eggs, hatch in the early spring across North America, just as leaves begin to emerge on the trees and shrubs that, ten months ago, moths laid these egg cases on. Early hatching caterpillars might have easier access to the new leaves on which they feed, *if* the leaves have already emerged. If leaves have *not* yet emerged, then early hatching caterpillars may not live long enough to find food. Late-hatching caterpillars, on the other hand, might hatch after all the new leaves have already been eaten, and thus may starve. Or they might hatch just when the leaves are emerging, after the early hatchers have already died, and may thus have plenty to eat. Or they might hatch after the leaves have already grown too big and too tough for these tiny caterpillars to chew through. A nonadaptationist approach to studying tent caterpillars that is rooted in contingency would simply accept that the fates of these caterpillars depend upon the vagaries of the timing of leaf emergence, which may or may not be influenced by the same factors that influence caterpillar hatching. Searching for these factors may well be interesting to some nonadaptationist biologists. But their search would be an open-ended one, without expectations for how the timing of leaf emergence and caterpillar hatching *should* be related in order to optimize fitness.

Furthermore, the fact that history unfolds contingently through time complicates our sense of causality. When everything depends on everything, and everything is always changing, how can scientists possibly figure out what causes any particular thing? The answer here is not to throw up one's hands in defeat because it no longer makes sense to search for a single, "true" cause of any particular phenomenon. Rather, any phenomenon could be explained by a multitude of possible causal pathways.[31] For example, an adaptationist narrative would claim that male elk have large antlers *because* males with bigger antlers are able to "dominate" males with smaller antlers, and, in turn, attract more females (an all-too-familiar narrative reminiscent of the mainstream description of Coho salmon, not to mention Victorian gender norms; see chapters 1 and 2). A more expansive nonadaptationist view, however, would suggest that the antlers of an elk are large and heavy possibly because of the antlers' role in a diversity of social interactions, *and also* because of how an elk's diet affects antler formation, *and also* because of the specific genetic pathways that shape antler development across an elk's lifetime, and also, and also, and also—all these causal pathways are *simultaneously* valid. Scientists can focus on how the material properties of elk antlers constrain their size and weight, and accept this as an always-partial explanation for why elk antlers are the way they are. An explanation that calculates the correlation of antler size and weight with survival and reproduction might be just as valid, but is also just as partial. These many causal pathways explaining a biological trait are interdependent, and so no single pathway can stand on its own as a "more true" cause of the trait.

The shift away from adaptationism toward co-constitution and contingency doesn't just affect how scientists think about evolution and causality. It also demands that scientists change how they think about the motivations underlying animals' actions. Under adaptationism, all animals are treated as rational yet unthinking agents, with a single, singular purpose for all their behavior: fitness optimization. Moving away from adaptationism therefore requires that biologists cease thinking about animals as rational yet unthinking agents alone, and instead expand their conception of what it can mean for an animal to truly have agency.

The history of animal behavior science reveals that the field has long grappled with this question of animal agency and the related question of anthropomorphism, which can be defined as the tendency to attribute human features to something nonhuman, including animals.[32] At present, by assuming that humans and animals alike behave as rational agents, the paradigms of optimality thinking and adaptationism restrict animal behavior science to a very specific kind of economic anthropomorphism. But this particular outlook on agency and anthropomorphism only arose during the ascent of sociobiology in the 1970s—it was not always the norm.

In the mid-nineteenth century to the early twentieth century, scientists were much more expansive and often entirely unabashed in their anthropomorphism, readily projecting a variety of their own social customs and norms onto animals (Darwin's projection of Victorian-era gender norms onto animals, as discussed in chapter 2, can be seen as part of this tradition). Consider the following scene of scorpion "courtship," written by the French naturalist Jean Henri Fabre, which shows how naturalists depicted animals as individual subjects who act with intention.

> 25th April, 1904.—Hullo! What is this, something I have not yet seen? My eyes, ever on the watch, look upon the affair for the first time. Two Scorpions face each other, with claws outstretched and fingers clasped. It is a question of a friendly grasp of the hand and not the prelude to a battle, for the two

partners are behaving to each other in the most peaceful way. There is one of either sex. One is paunchy and browner than the other: this is the female; the other is comparatively slim and pale: this is the male. With their tails prettily curved, the couple stroll with measured steps along the pane. The male is ahead and walks backwards, without jolt or jerk, without any resistance to overcome. The female follows obediently, clasped by her fingertips and face-to-face with her leader.

The stroll is interrupted by halts that do not affect the method of conjunction; it is resumed, now here, now there, from end to end of the enclosure. Nothing shows the object which the strollers have in view. They loiter, they dawdle, they most certainly exchange ogling glances. Even so in my village, on Sundays, after vespers, do the youth of both sexes saunter along the hedges, every Jack with his Jill.

Often they tack about. It is always the male who decides which fresh direction the pair shall take. Without releasing her hands, he turns gracefully to the left or right about and places himself side by side with his companion. Then, for a moment, with tail laid flat, he strokes her spine. The other stands motionless, impassive . . .

At last, about ten o'clock something happens. The male has hit upon a potsherd whose shelter seems to suit him. He releases his companion with one hand, with one alone, and continuing to hold her with the other, he scratches with his legs and sweeps with his tail. A grotto opens. He enters and, slowly, without violence, drags the patient Scorpioness after him. Soon both have disappeared. A plug of sand closes the dwelling. The couple are at home.[33]

Subsequently, in the early to mid-twentieth century, many animal behavior scientists (also called *ethologists* in this era) tried to distance themselves from the thorny problem of agency and anthropomorphism altogether, by focusing on those behaviors of animals that they considered "instinctual"—fixed and somewhat robotic—and using only the most technical and precise language, devoid of feelings or desires, to describe these behaviors. But the ethologists' technical language only disguised, rather than deleted, their anthropomorphism—all human language is *necessarily* anthropomorphic because language is how humans understand one another. For example, science historian Eileen Crist quoted ethologist Konrad Lorenz's description of a goose: "If I am walking along with a tame greylag goose which suddenly stretches, extends its neck and softly utters a harsh warning-call, I may say 'now it is alarmed.' However, this subjective abbreviation only means that the goose has perceived a flight-eliciting stimulus."[34] Despite ethologists' attempts to erase all human idioms from their descriptions of animal behavior, and even though Lorenz thought it was inappropriate ("this subjective abbreviation") to use the term "alarmed" in his description of a goose, he still had to describe this goose as "alarmed" so that his readers could better understand what it looks like when a goose "has perceived a flight-eliciting stimulus."

Moreover, by favoring technical and precise language, these early to mid-twentieth-century biologists inadvertently began engaging in anthropodenial. Coined by animal cognition scientist Frans de Waal, *anthropodenial* refers to the tendency to avoid using similar language to describe similar behaviors in humans and animals, in the name of scientific objectivity. This tendency persists in mainstream animal behavior science today. But are scientists really being more objective when they describe how apes respond to being tickled, for example, as making "breathy sounds with a rhythm of inhalation and exhalation," rather than simply as "laughing"?[35] Equally, are scientists who engage in anthropodenial in fact failing to appreciate animal behaviors fully when they deliberately ignore animal behaviors' similarities with human behaviors? In short, this history reveals that anthropomorphism is both inevitable and not necessarily a problem for animal behavior science; what *is* a problem, however, is the failure to examine the entanglements of the field's anthropomorphism, and its conceptions of agency, with human systems of power.

Viewing animals as having a broad-ranging agency does not mean believing that animals are infinitely capable of doing whatever they please, nor that they necessarily have a consciousness akin to humans, nor that their decision-making is exactly like ours.[36] It simply means that they are capable of acting in complex, undeterminable, unpredictable ways for any number of reasons. With an expansive view of animal agency, animals' behaviors can no longer be assumed to serve a single purpose, and animal behavior scientists cannot know or prove that a particular behavior exists *for* a particular purpose. Maybe animals of all kinds engage in social and sexual interactions because these interactions can be pleasurable. Maybe squirrels gather acorns and chickadees cache seeds to stave off boredom, or to soothe an anxious mind. Maybe monkeys throw things at their human handlers from a deep sense of injustice.[37] Maybe elephant seals rescue their friends' pups who have swum too far out to sea simply because they care about one another.[38] Maybe that one crow slides down a snowy roof on a plastic lid, over and over again, because he once saw another crow do the same thing and it looked thrilling and so he tried it and it was.[39] Maybe animals sometimes do things (or sometimes do absolutely nothing) *just because* ¯_(ツ)_/¯.

Indeed, organism-environment co-constitution depends upon organisms being the opposite of passive, and having the agency to change the conditions of their own lives. Agency also allows for contingency, which historian Jessica Riskin has described as "neither randomness nor determinism but something altogether different. It is the product of limited agencies working in particular changing situations."[40] Moving away from adaptationism in animal behavior science, and toward co-constitution and contingency, thus depends upon believing that animals have agency. While you might be perturbed by our use of the word "believe," consider that allowing for a worldview in which animals have agency is no more or less of a belief (or a political, ethical, and spiritual commitment) than limiting animal agency, or denying it altogether.

Limiting and denying agency is central to another system of oppression that dominated both scientific and political worlds in the late nineteenth century and first half of the twentieth century: eugenics. In the next two chapters, we discuss how eugenics remains deeply entangled with multiple facets of mainstream animal behavior science, and how the field might extricate itself from this entanglement.

Figure 6.1
A large copper dung beetle on a ball of dung. Original in color. Photo: Bernard Dupont.

6 EUGENIC ENTANGLEMENTS

THE FALLACY OF "SMARTER"

As their name suggests, dung beetles live a life that is, well, all about dung. Found across the world on every continent except Antarctica, these iridescent, jewel-like insects are impressively steadfast—they are born in dung, eat dung their whole lives, and lay their eggs in dung. And some species of dung beetle go so far as to make a little dung world of their own. Descending in droves upon a pile of poo, individual beetles each quickly roll away a portion, and then keep rolling. The beetle pushes his dung ball backward—no mean feat, since the dung balls are sometimes bigger than the beetles themselves. To push, the beetles position their bodies at what seems like a precarious angle: hind legs high up on the dung ball, fore legs tiptoeing on the ground, face pointed almost directly downward.[1]

As the beetles roll their dung balls, in search of places to store them, they must maneuver across the ground. A grassland or field might offer easy, open paths to us, but to a dung beetle, these landscapes are a veritable jungle. How do dung beetles keep track of where they're going? Every now and then, they pause their rolling and climb up on top of the dung ball (figure 6.1), look around, sometimes change direction, then begin pushing again. Researchers have learned that when dung beetles climb on top of their dung balls and look around, they are reorienting themselves by tracking the position of the sun.[2] And darkness doesn't stop them either: at night, they orient themselves by the stars of the Milky Way.[3] Dung beetles are celestial beings!

It's awe-inspiring to think of the humans who, many centuries ago, were able to find their way across vast swathes of land and ocean using the sun and the stars as their guides.[4] But to think of a dung beetle doing something similar? It almost beggars belief—this remarkable capacity for sensory perception and cognitive processing, a process comparable to some of the most intrepid human voyagers, in a *beetle*?

Cognitive ethologist Frans de Waal might have argued that we're foolish to question dung beetles' capabilities. Rather, we should question our surprise. We can be amazed, yes, by the feat of celestial navigation itself, but not because it's a beetle accomplishing the feat. De Waal argued in his 2016 book *Are We Smart Enough to Know How Smart Animals Are?* that all animal species are as smart as they need to be to make their way through their worlds. And because different animals live in such different worlds, they're all smart in different ways. "We had better use the plural to refer to their capacities, therefore, and speak of intelligence*s* and cognition*s*," de Waal wrote.[5]

De Waal's claim runs counter to the pervasive, insidious tendency to arrange animals into a hierarchy of sophistication—a tendency that can be traced back several millennia to Aristotle's *scala naturae* and the medieval Christian notion of the "great chain of being." Both of these frameworks ordered all manner of beings linearly, from the divine at the very top, to angels, to humans, to other mammals and "lower" animals, and on downward to plants and minerals at the very bottom (just above hell). It's this hierarchy that can lead us, even unconsciously, to be surprised by so-called lesser animals doing things as complex as navigating by the stars (dung beetles), communicating detailed information through dance (honey bees), or picking out their own offspring's calls from the cacophony of hundreds or even thousands (many species of bats).[6] And these expectations of hierarchy have only hindered scientists' ability to appreciate animal cognition on the *animals'* terms. As de Waal wrote: "Comparisons up and down this vast ladder have been a popular pastime of cognitive science, but I cannot think of a single profound insight it has yielded. All it has done is make us measure animals by human standards, thus ignoring the immense variation in organisms' *Umwelten* [the world that an animal perceives through all their senses; see chapter 2]. It seems highly unfair to ask if a squirrel can count to ten if counting is not really what a squirrel's life is about."[7]

Yet media reports on animal cognition are still published, in 2023, with headlines like "The 30 Most Intelligent Animals in the World Might Surprise You" (unsurprisingly, humans are placed at the top of that list).[8] And the authors of the *Animal Behavior* textbook end their chapter on neuroscience and animal cognition by posing this question as one worth investigating: "What exactly is it about human brains that contribute to our superior intelligence?"[9] Despite de Waal's persuasive argument to the contrary, animal behavior science remains in the grip of hierarchical thinking about cognition.

Stepping back from animal cognition, we can recognize that the hierarchical ideas of the *scala naturae* and the great chain of being run deep across much of dominant animal behavior science and evolutionary biology. Prior to and alongside Darwin's formulation of evolution by natural selection, hierarchical thinking was foundational to the exercise of classifying living beings into different species. This exercise was, in turn, intimately tied up with powerful Europeans' preoccupations with establishing their own superiority and domination over the poor and the disabled, over women and queer people, and over people of color and people from other continents. Scientists at the time largely belonged to privileged rather than marginalized groups, and so the hierarchies that scientists constructed to make sense of the natural world were deeply entangled with the hierarchies that undergird human systems of oppression. These entanglements, and their specific connections to eugenics, are illustrated perfectly by the life and work of turn of the twentieth-century American fish taxonomist David Starr Jordan (1851–1931).

Jordan had a childhood fascination with inspecting and trying to make sense of the living world. This fascination turned into a career when he began studying under the zoologist and geologist Louis Agassiz.[10] Agassiz's motivations in paying attention to nature went beyond mere fascination, however—he wanted to understand how God had ordered the world while creating it. In her book *Why Fish Don't Exist*, journalist Lulu Miller described how "Agassiz believed that hiding in nature was a divine hierarchy of God's creations that, if gleaned, would provide moral instruction. . . . Agassiz believed that by arranging these organisms into their proper order, one could come to discern not just the intent of a holy maker but also instructions for how to become better."[11] Darwin's theory of evolution directly challenged the idea that all living species had been created by God and, unlike Agassiz, Jordan agreed with Darwin. Nonetheless, Jordan retained his teacher's commitment to hierarchy. As Miller put it, Jordan "was still on the hunt for the shape of the ladder that revealed how all creatures and plants were ordered—only now he believed its arrangement had been forged by time, not God."[12]

For both Agassiz and Jordan, this commitment to a hierarchical sense of nature was reflected in their virulently racist and ableist descriptions of human variation. Moreover, their scientific approach to categorizing and ranking animals lent credence to their political claims about how humans must be categorized and ranked, both racially and by physical and cognitive "ability." For example, Agassiz ranked human racial categories hierarchically, and continued to endorse the theory of polygenism (the theory that different human "races" are wholly different species) long after Darwin theorized that all humans descended from a single, primate ancestor (a theory called monogenism).[13] And in his 1907 book called *The Human Harvest: A Study of the Decay of Races through the Survival of the Unfit*, Jordan described people with disabilities as "human beings with less intelligence than the goose, with less decency than the pig," and proclaimed that it would be a true act of charity to "guarantee that each individual *crétin* should be the last of his generation."[14] Jordan's claim to scientific understanding about the evolutionary history of animals gave his predictions about the future evolution of humankind an air of authority, and established him as one of the foremost proponents of eugenics in the US. In fact, the eugenics movement, which sought to control the future of humanity by controlling who gets to reproduce and who does not, was intimately entangled with the work of evolutionary biologists more generally.

The eugenics movement was not some fringe exercise nor restricted to Nazi Germany but rather a widespread ideology across the US and Europe from the mid- to late nineteenth century and well into the twentieth. President Theodore Roosevelt, for example, was a prominent proponent of eugenics.[15] The dominance of eugenic ideology was bolstered by scientific authority; Nazi Germany's brutal extermination programs are widely documented to have been inspired by US-based eugenics scientists.[16] Inquiry in the realm of evolutionary biology was especially relevant and consequential. Francis Galton (1822–1911), the founder of eugenics and a cousin of Charles Darwin, explained the connection clearly. According to Galton, it was incumbent upon humans to intervene in the process of evolution for the betterment of the species. "We may not be able to originate, but we can guide," Galton wrote. "The processes of evolution are in constant and spontaneous activity, some towards the bad, some towards the good. Our part is to watch for opportunities to intervene by checking the former and giving free play to the latter."[17] What traits, and which people, the eugenicists considered "good" or "bad" were shaped by interlocking systems of power. The tools of eugenics—programs of forced sterilization and other forms of population control, institutionalization, and genocide—were thus deployed against people marginalized on the basis of race, gender, disability, sexuality, and class.[18]

Jordan's specific eugenic fantasies came to intersect with a scientific and economic project begun by railroad mogul and university founder Leland Stanford: the project of breeding racehorses. In *Palo Alto*, journalist Malcolm Harris described how Stanford's system for breeding horses served as an important driver of capitalist expansion in California and the US. "For Stanford the capitalist," Harris wrote,

> the horses were productive biological machines, and in races he could analyze their output according to simple, univocal metrics . . . faster horses were better horses, and if he could master the production of better horses, then he could improve the country's capital stock. Stanford figured that if through the application of scientific methods he could raise the value of the average horse by $100, that would be worth $1.3 billion to a country with 13 million horses (over $30 billion in 2022 money).[19]

This example makes clear how readily the mechanics of capitalism can be fueled by the mechanics of artificial selection.

Jordan, who became Stanford University's first president, saw direct parallels between his employer's largely economic project and his own human eugenic goals—the mechanics of capitalism and artificial selection are also the mechanics of eugenics and adaptation by natural selection. "In the animal world," Jordan wrote, "permanent progress comes mainly through selection, natural or artificial, the survival of the fittest to become the parent of the new generation. In the world of man similar causes produce similar results."[20] At Stanford, Jordan surrounded himself with men who were also motivated to leverage their scientific authority to establish supremacies on the basis of race, ability, and intelligence. One of them, professor of education Lewis Terman, developed the metric that became foundational to Jordan's eugenic efforts, the human equivalent of horse speed, if you will: intelligence quotient, better known as "IQ."[21] Armed with this metric, Harris argued, the likes of Jordan and Terman established Stanford University as "a positive eugenic project, breeding high-IQ people to produce the next generation of Palo Alto residents."[22]

This eugenical vision of the world was not just racist and ableist but also patriarchal; women, in this vision, were simply machines that produced men's children. As Harris put it, "in this division of labor, men pursue their interests as individuals while women consolidate men's achievements for the species—or, more precisely, the race."[23] Given the primacy of reproductive control—who mates with whom, and who does not—to the eugenic project, it is not surprising that the science of sexual interactions became deeply intertwined with eugenic thinking. These entanglements persist in mainstream animal behavior science to this day, in its theories of sexual selection.

We return to the subjects of intelligence hierarchies and their intertwining connections with racism, ableism, classism, and eugenics in the next chapter. Here, we explore the striking parallels between theories of sexual selection and eugenic thought.

WHITHER SEXUAL SELECTION?

In 1871, Darwin proposed a theory of *sexual* selection (distinct from his earlier theory of *natural* selection) to explain the huge diversity of biological features that do not appear to help organisms survive and therefore could not have arisen by natural selection. If anything, features like bright feathers on birds, heavy antlers on elk, and bizarrely shaped appendages protruding from the heads of treehoppers are more likely to *hinder* survival than help it, so why do they persist in the face of natural selection's relentless optimization for survival? Darwin's answer to this dilemma remained within the realm of adaptationism and optimality thinking; he proposed that such features must instead contribute to animals' fitness by ensuring their mating success. By functioning as *armaments* or *ornaments*, these structures must either help the animal win fights against competitors or help the animal attract mates (or both).

But why exactly would a female peahen choose to mate with a male peacock with a gigantic tail that likely slows him down, and only increases the odds that he will be eaten by a predator? Specific hypotheses abound, all of which tend to fall into one of two types (with both types assuming that the appearance of a peacock's tail is encoded in his genes). According to the first type of hypotheses, the "good genes" hypotheses, the large flashy tail indicates the male's superior genetics and capacity to thrive in the face of resource scarcity, competition, predation, or disease, notwithstanding the hindering presence of the tail. According to the second type of hypotheses, known as "runaway" hypotheses, the large flashy tail is simply an aesthetic marvel, and thus irresistible to the female's eye for beauty, despite its costs.[24] Both of these hypothesized mechanisms of sexual selection are expected to lead, over many generations, to the evolution of ornaments and armaments across animals, notwithstanding their presumed negative effects on animals' survival. These two mechanisms continue to dominate explanations in mainstream animal behavior science for animals' showiest, most unwieldy traits. Diving into these mechanisms in depth, and situating them in historical context, reveals them to be inextricably linked with eugenics.

That some individual animals happen to mate or reproduce more than others is, in and of itself, not necessarily aligned with eugenic thinking. But the logic of the hypothesized mechanisms of sexual selection—both the good genes and the runaway hypotheses—goes beyond that: this logic enlists individual animals, females in particular, into doing the work of *choosing* who gets to reproduce and who does not, based on perceivable indicators of males' superior or inferior genes (whether for pragmatic or aesthetic reasons).

This is the same logic that eugenicists apply to human populations through what's called "positive eugenics," which means encouraging those people with "favorable" traits to breed. It's not as though this parallel is particularly hidden from view: the word "eugenics" is derived from the Greek *eugenes*, which means "good in birth" or "good in stock," or, as a scientist studying sexual selection might put it, "good genes." And the mechanism of runaway sexual selection was first described by Ronald Fisher, an evolutionary biologist and committed eugenicist who "mathematized evolutionary theory so that it was consistent with his eugenic hopes and fears for human evolution."[25]

Sexual selection theory was developed during the heyday of eugenics in the late 1800s and early 1900s, a time at which (as we've seen in chapters 2 and 3) biologists had no qualms at all about developing scientific theories that directly supported their political agendas. Sexual selection theories in particular were used to bolster both eugenic panic (fear about what might happen to human society if nobody intervened in people's mating choices) as well as eugenic controls (*how* to intervene). As historian Erika Milam wrote in her book on the early history of sexual selection, *Looking for a Few Good Males*: "Sexual selection . . . served both as a viable method of eugenic improvement and as a cautionary tale about the inevitable effects of the fertility trends of modern society. As long as the preeminent members of society were more fertile than less desirable members, and women continued to choose their husbands well, then society would become more intelligent and more beautiful."[26]

Leonard Darwin, son of Charles Darwin and an avid eugenicist, made these connections between sexual selection and eugenic control abundantly clear at the Annual Meeting of the Eugenics Education Society in 1923, an organization for which he served as president.[27] In his presidential address, the younger Darwin opened his speech thus: "Wonderful results have been produced . . . by the action of sexual selection in regard to the evolution of all kinds of organisms which are endowed with the power of sense perception; and, if this be so, ought we not to enquire whether this same agency cannot be utilized in our efforts to improve the human race?"[28] He then went on to deliver an astoundingly modern-sounding description of various mechanisms of sexual selection before moving to consider the implications of these mechanisms for the most effective eugenics propaganda, urging people to make better mate choices. Leonard Darwin believed that: "To secure human progress, the inferior types must be eliminated; and all that should be demanded is that this process should be made as little painful as possible and that liberty should never be unnecessarily limited."[29] Sexual selection offered him and his fellow eugenicists a scientific foundation upon which to develop and enact this "less cruel" approach.

The good genes hypotheses for the evolution of animal traits are directly analogous to positive eugenics, as well as being entirely consistent with optimality thinking (which is not a coincidence but rather reflects how eugenics and capitalism feed into one another). For example, a female peahen might choose to mate with the peacock with the brightest feathers because feather brightness is correlated with, say, the ability to find the best feeding patches, and is therefore correlated with fitness. Under this good genes scenario, choosing a mate with bright feathers is rational and adaptive—her offspring will have high fitness thanks to having inherited the capacity to find feeding patches from the male that sired them. These matings will propagate desirable traits such as optimal patch-finding ability in future generations of peafowl, and will lead to the betterment of the species, which eugenicists would consider a favorable outcome.

In contrast to the direct eugenic implications of good genes hypotheses, the eugenic entanglements of the runaway mechanism are more complex. According to runaway sexual selection, a peahen might mate with the male with bright feathers because she's attracted to those feathers for no rational reason at all—an arbitrary aesthetic choice that may in fact be detrimental to fitness. This decoupling of desirability from fitness may seem antithetical to eugenics—how could such a selective process lead to the betterment of the species? It turns out that runaway sexual selection is exactly aligned with eugenic goals, not least because Fisher intended it to be used as a tool of eugenics.

Runaway sexual selection was Fisher's solution to a strange incoherence that sits at the heart of the eugenic project. On the one hand, according to the eugenicists, the traits that distinguish wealthy, white Euro-Americans were obviously superior and worthy of being preserved. On the other hand, eugenicists believed that without their interventions, wealthy, white Euro-Americans were in constant danger of being outnumbered by the poor, Black, and Brown masses who were seen as *too* successful at survival and reproduction despite their presumed inferiority.[30] Thus the eugenicists were especially concerned with exactly those traits that they saw as desirable—preserving racial and class purity—but that didn't naturally promote individual survival or reproductive success.[31] Fisher and his ilk therefore believed that society needed the help of eugenics to ensure that the most deserving traits persisted and spread. For example, Fisher cared deeply about selecting for heroism in men, or the willingness to sacrifice oneself in battle.[32] Being heroic in battle is unlikely to benefit survival, but Fisher was hopeful that social norms that made heroic men returning from war into attractive marriage partners could nonetheless increase the prevalence of heroism in certain human populations. Fisher's theory of runaway sexual selection offered the eugenicists a way to defy nature by controlling reproduction, all to achieve their desired end of domination.

Mainstream animal behavior science has yet to reckon in any meaningful way with the deep ties between sexual selection theory and eugenics—good genes hypotheses and runaway hypotheses remain at the center of how science makes sense of animals' sexual behavior. As we've seen in earlier chapters, more expansive conceptions of animal agency coupled with queer standpoints on animal interactions offer promising starting points from which to reimagine the science of animal sex, and animal behavior more broadly. But in order to fully extract ourselves from mainstream animal behavior science's eugenic entanglements, we must tackle a final paradigm that is central to how hierarchies are naturalized: the paradigm of biological determinism.

Figure 7.1
A female lance-tailed manakin watches while, behind her, two male lance-tailed manakins display together. Photo: Emily DuVal.

7 DISMANTLING DETERMINISM

MATING IN MANAKINS

When she began her research in 1999, behavioral ecologist Emily DuVal did not intend to study how female lance-tailed manakins choose their mates; she was interested in cooperative interactions between males during their delightful singing and dancing displays (we discussed DuVal's findings on male cooperation in chapter 3). While studying the male birds' displays and reproduction, DuVal collected all the data she could find on male behavior and morphology, and also individually tagged and followed the females to measure how many eggs they laid and how many of their chicks survived, as a way of calculating male fitness. DuVal soon realized that she could use these exact same data to explain how females choose which males to mate with.

The possibilities for how female lance-tailed manakins might choose their mates seemed endless. Were females choosing something about the males' dancing and singing—perhaps how long the pairs of males went on, or how coordinated they were? Was it something about the males' features—the brightness of their blue shoulders, or the length of their (admittedly quite stubby) tails? DuVal and her colleagues performed every analysis of mate choice and sexual selection they could think of, from both good genes and runaway perspectives. But, over and over again, the papers they published from these mainstream perspectives showed inconclusive results. A male's age, experience, and genetic makeup were somewhat correlated with whether and how often he mated with females, but, by and large, the only consistent finding was variation.[1] Different females "were assessing the same group of males and making different choices," DuVal told us.[2] "It's now twenty-five years of studying these birds, and I still don't know what females are picking when they choose their mates. . . . It's not the easy answer I thought I would potentially find when I dug into what the females are doing. It's really complicated."

DuVal's bafflement at the tremendous variation in female lance-tailed manakins' mating choices is yet another moment of a paradigm being confronted with an anomaly. This time, the paradigm in question is that of biological determinism, which insists that the traits underlying all kinds of hierarchies in humans and animals are somehow locked into our biology. Narratives of biological determinism are common in scientific accounts of why animals behave as they do. You've no doubt seen the logic of biological determinism, perhaps in headlines like "Why Are Dogs So Friendly? The Answer May Be in 2 Genes" or a more poetically titled piece about deer mice burrows: "The Genes That Built a Home."[3] The standard models of sexual selection assume biological determinism—in order for female animals' mating preferences to evolve by natural selection, these theories posit, their preferences must be genetically determined.

Twenty-five years of assuming female's mate preferences to be genetically determined hadn't gotten her very far in understanding them, so DuVal began wondering whether a female's mate preference might instead emerge from her experiences, specifically her social interactions. This would mean that female mate choice is both *contingent upon* and *co-constituted with* her social and ecological environment. DuVal subsequently demonstrated, along with behavioral ecologist and evolutionary biologist colleagues Courtney Fitzpatrick, Elizabeth Hobson, and Maria Servedio, that female mate choice can be understood, in all its glorious variability, as the outcome of females *learning* which males to choose by watching other females, but learning *imperfectly*.[4]

DuVal and colleagues explored their ideas about female choice and social learning with a mathematical model. This model can apply to any situation where individual animals can watch others choose their mates (from fruit flies to guppies to sage grouse). But it's easiest to understand this model by imagining how it would work in lance-tailed manakins. Pairs of male lance-tailed manakins dance and sing in clearings across the forest understory, and females move among different pairs, eventually choosing one mate from a given pair. Let's say that the particular female we're watching flies among five different pairs of males and then chooses to mate with the male with the longest tail. Another female has been following her quietly, watching all her interactions as well as her ultimate choice, to try and infer how one might choose a mate. Let's assume that the second female cannot simply ask the first female to explain her choice. Instead, based on what she's learned from watching the first female, the second female makes an informed guess as to how to choose a mate.

Now imagine that, in this particular set of ten males, the male with the longest tail also happened to have the bluest shoulders. The second female happens not to notice tail lengths, but instead thinks, "Aha! The bluest shoulders! That's how I choose a mate!" The second female thus goes on to a lifetime of choosing to mate with the bluest-shouldered males, even though she learned her mate choice from a female who was actually choosing the longest-tailed males. More generally, depending on the subset of males they encounter while learning, the females who are watching might sometimes end up choosing the same traits as their teachers, and sometimes not. When this mechanism is extended to the whole population, one *expects* that females will vary in which mates they choose; conformity does arise sometimes, but only in very particular ecological and social circumstances. Essentially, DuVal and colleagues' model shows that moving away from determinism and accounting for the contingencies of a female's life and the forces of co-constitution—where she happens to roam, whom she happens to interact with, and what she happens to learn—leads one to *expect* variation among females in their choice of mate.

When we make room for the vagaries of an animal's life, our understanding of animal behavior expands. Expecting that animals' lives unfold in unpredictable, agential, and relational ways allows us to turn decisively away from the rigid hierarchies and normative expectations that are central to theories of animal sex and that are rooted in the twin ideologies of eugenics and biological determinism. These ideologies together have long been used to enact harm and violence against people marginalized on the basis of sex, race, class, and ability. Again, mainstream animal behavior science has been central to justifying such harms.

RACE, EUGENICS, AND BEHAVIORAL GENETICS

In his book *Misbehaving Science*, sociologist Aaron Panofsky wrote about the eugenic origins of the field of behavioral genetics established in the early twentieth century. He noted that the field's founding scientists were "motivated by the desire to improve the gene pool," which entailed proliferating those genes associated with positive social behaviors and eliminating those genes assigned to negative behaviors, including "feeblemindedness, imbecility, pauperism, criminality, prostitution, alcoholism, and the like." Which traits were categorized as positive or negative depended, of course, on the social locations of the scientists doing the categorizing. As Panofsky wrote: "To many privileged social groups the seeming concentration of these traits among the growing laboring classes, immigrants, darker races, and inhabitants of urban ghettos represented a looming danger. For many elites, such traits—as well as their moral obverses, eminence and genius—had long been considered organic, immutable, and hereditary. The emerging science of genetics promised to rationalize these sentiments."[5] Genetics offered a twentieth-century tool that could be leveraged to insist that existing hierarchies were largely unchangeable except through eugenics. But the scientific project of ranking people in order to justify domination and oppression has a history much older than genetics. White Europeans throughout the sixteenth to the twentieth centuries rationalized their violent treatment, dispossession, enslavement, and colonization of peoples from the African continent, South Asia, Australasia, and the Americas by considering racialized peoples as separate and "naturally" lesser species (polygenism). During the mid- to late nineteenth century, as Europeans began to more widely accept that all humans belonged to the same species (monogenism), they could no longer justify their behavior toward racialized peoples through species rankings. Rather than cease their violent—and profitable—enslaving and colonizing missions, elite Europeans and Euro-Americans simply morphed their polygenist hierarchy of human species into a monogenist hierarchy of human races.[6]

What did these European naturalists and scientists base their racial hierarchies upon? Historian Peter Bowler described how "the explorers of the Victorian era routinely pictured the peoples they encountered as mentally and morally less advanced than themselves."[7] Moreover, they attempted to tie these more abstract differences to observable, biological ones; according to Bowler, "As Europeans began to conquer, enslave, and even exterminate other races, there was a tendency to exaggerate these racial differences to justify the exploitation. If the nonwhite races were less than fully human, it was easier for whites to feel comfortable with a situation in which the superior race determined the fate of the inferior."[8] They ranked people of African descent as lowest, and therefore most similar to our animal ancestors, while ranking themselves, unsurprisingly, highest, and therefore most superior to animals (ordering other human "races" somewhere between these two poles). Some nineteenth-century European and Euro-American naturalists and explorers went so far as to kidnap various people of color for "display" in zoos alongside stolen animals, further underscoring non-Europeans' supposedly "savage" nature and animality.[9] As Claire Jean Kim, a contemporary scholar of political science and Asian American studies, observed: "Race has been articulated in part as *a metric of animality*, as a classification system that orders human bodies according to how animal they are—and how human they are not."[10]

Given this historical context, it's not surprising that once the tools of genetics were developed, they were employed in service of the idea that humans' mental and moral characteristics were genetically determined and correlated with race. From the 1969 article by Arthur Jensen titled "How Much Can We Boost IQ and Scholastic Achievement?" to the 1994 book by Charles Murray and Richard Herrnstein, *The Bell Curve*, many modern scientists have argued that differences in educational and economic success among people categorized as different races can be explained by differences in intelligence, which in turn can be explained by differences in genetics. These scientists contended, as Panofsky put it, that "discrimination is mostly over, and that unequal social structure is genetically determined. Policies aiming to uplift minorities and the poor are [therefore] doomed to fail . . . instead, the 'cognitive elite' must find ways to manage a permanent genetic underclass."[11] In short: genetic determinism cements inequality into our biology, and is thus used to rationalize political inaction in response to inequity. Worse, ideas of genetic determinism continue to bolster white supremacist ideology. In his rambling manifesto, the white perpetrator of a 2022 mass shooting targeting Black people in Buffalo, New York, cited recent research on relationships between genes and educational attainment to justify his actions.[12]

Thankfully, there has also been a long history of scholars, including biologists, who have demonstrated that racial categorizations do not map neatly onto human biological differences, genetic or otherwise, and who do not agree that it's possible to isolate the genetic "causes" of a trait, especially not one as nebulously defined as intelligence.[13] For example, in 1974, Richard Lewontin described how "a causal pathway may go from tryptophane metabolism to melanin deposition to skin color to hiring discrimination to lower income," but a naïve biologically deterministic analysis would nonetheless attribute a genetic cause to something like "economic success."[14] Which is to say, it's certainly *possible* to characterize economic success as being genetically caused. But to do so is to ignore all the cultural and social systems (of oppression) that invest a biological trait like skin color with economic and political meaning. Or, as feminist biologist Ruth Hubbard put it, in 1993: "We cannot sort nature from nurture when we confront group differences in societies in which people from different races, classes, and sexes do not have equal access to resources and power, and therefore live in different environments."[15] Or, as a 2021 team made up of human geneticist Brenna Henn, science studies scholar Emily Klancher Merchant, geneticist and anthropologist Anne O'Connor, and philosopher Tina Rulli described:

> Our problem with . . . [the] focus on genes as the primary cause of social inequality is that it elides the role of power. It presents our existing social hierarchy as the result of (genetically driven) success or failure at the individual level, neglecting the policies and practices that systematically benefit some types of people (such as employers, landlords, and creditors) at the expense of others (such as employees, tenants, and borrowers). It renders invisible the ways in which the wealthiest members of our society get ahead through the active exploitation of the most vulnerable.[16]

The long history of fierce opposition to biological determinism highlights both how stubbornly entangled this paradigm is with seemingly every system of oppression and how deeply entrenched it is in mainstream animal behavior science. Rooting it out might seem impossible. But it's worth recognizing that, in a way, we've already confronted determinism several times earlier in this book. We encountered a kind of determinism in Darwin, Bateman, and Trivers's patriarchal claim that female animals are inherently coy and choosy, and we broke down this determinism by focusing on the contingent unfolding of females' lives. We faced a kind of determinism in the sociobiologists' capitalist belief that animals respond to their environments in a way that is inherently rational and optimizing, and we broke down that determinism by focusing on how all animals' lives are co-constituted with their environments. More broadly, the dual paradigm of adaptationism and optimality thinking is also a kind of determinism entangled with every system of oppression—under this paradigm, any trait *will* be optimized by natural selection, and it *will* become an adaptation if it isn't one already, which means that its fate is determined. And our route out of adaptationism and optimality thinking is, once again, a focus on contingency and co-constitution.

All of this is to say that we already know how to break out of the paradigm of biological determinism—with contingency and co-constitution. We've already seen an example of this too: as we considered female mate choice in lance-tailed manakins previously in this chapter, contingency and co-constitution offered us an alternative to the genetic determinism of females' mating preferences that is assumed in mainstream theories of sexual selection. In the following section, we offer three examples—one of two humans, one of a human-animal interaction, and one that is about all of us humans and animals—that together illustrate how co-constitution and contingency can dismantle biological determinism once and for all.

BRADLEY AND BOGUES

Behavioral geneticist and self-identified progressive Kathryn Paige Harden argued, in her 2021 book *The Genetic Lottery*, that persistent social inequities can only be remedied after accounting for their genetic underpinnings. Harden claimed that because at least some of the (presumed) root causes of social inequities and injustices are (presumed to be) biologically determined, political interventions to redress inequities and injustices cannot be effective unless they account for these (presumed) biological differences.

In his critical response to *The Genetic Lottery*, evolutionary biologist (and avid basketball fan) Brandon Ogbunu deconstructed Harden's logic by taking on one of her book's least controversial claims: that genes underlying human height can determine life outcomes, as exemplified by the very successful, seven-and-a-half-foot tall basketball player Shawn Bradley. With this example, Ogbunu explained, Harden had hoped to achieve the following goals: first, to demonstrate that, because his success depended on his height, Bradley won the genetic lottery when fate dealt him the specific combination of genes that "made" him tall; and second, to extrapolate wildly from this example to all kinds of people and professions, concluding more generally that "some of us are born with the tools that provide a predisposition to success, and [that] we can imagine another world, where we use knowledge of our genetic hauls to structure a more fair society."[17]

But, Ogbunu pointed out, Harden's argument failed because she did not appreciate how Bradley's success depended not only on the genes underlying his height but also, more foundationally, "on the lucky draw of being born and coming of age during a time when there is a professional sport (basketball was invented in 1891, and the NBA founded in 1946) that rewards especially tall, long-limbed individuals. One struggles to find any other sport (or profession, for that matter) where people standing 7′6″ have a distinct professional advantage, outside of basketball."[18] Bradley's success as a basketball player was thus contingent upon his height *as well as* the social and historical conditions into which he was born. We can also recognize that agency played a part here: Bradley's success at basketball likely depended, in no small part, on him repeatedly choosing to show up for practice, rather than, say, choosing to use that time to memorize Shakespeare monologues (we're guessing here; we don't know for a fact that Bradley wasn't also a theater kid).

And while being tall can certainly help one become a successful basketball player, not all tall people are good at basketball, and, more importantly, height is not the only route to success in this sport. Ogbunu contrasted Bradley's career with that of Bradley's contemporary Tyrone Bogues, who is five foot three. Bogues's success could be attributed not just to his stellar training (which is at least partially attributable to the luck of being born in a particular city) but also, again, to context—his successful strategies of "darting in and out of small spaces . . . zipping up and down the court nimbly and stealthily" depended on his being relatively short compared to his teammates and opponents.[19] To claim, in a biologically deterministic way, that genes for being tall cause success in basketball is to miss the most interesting part of the game.

And it's not just nebulous traits like "success," or human contexts like basketball, for which genetic determinism fails—the contribution of genes to a specific trait in an organism is *never* separable from the organism's environment. The length of a baby giraffe's neck, for example, may be correlated with the length of his parents' necks, and it would be tempting to infer that the similarities in the lengths of their necks are caused by their shared genes. But in fact that correlation may have little or nothing to do with the genes passed on from parent to offspring. Perhaps one parent's own long neck allowed her to search farther for new patches of savanna to move to, or to forage on leaves that weren't accessible to other giraffes in her herd, which meant more nutrition for her growing fetus, who, as a result of this abundance, ended up being born with a long neck too. Yes, the growth of his neck depended on certain genes being expressed in his developing body, and maybe the extra nutrition allowed those genes to be expressed to a greater degree, turning his neck even longer. His genes were undeniably *involved* in producing his neck in the first place. But does it really make sense to say his genes *caused* his neck to be longer? Or does it make more sense to say, instead, that the length of a baby giraffe's neck is an emergent result of contingent interactions between his genes and his environment (including his environment in utero), which are in turn shaped by his parents' biology and behavior, which are co-constituted with the conditions of the ecosystems they inhabit?

Imagine that the development of a giraffe's neck (or any other trait) is like filling a bucket. Imagine also that the bucket is being filled by two people, one labeled "genes" and the other "environment."[20] In a world of biological determinism, the two people would each have their own hose with which to fill the bucket, and one can imagine measuring the diameter of the hose as well as the rate of water flow, the volume of the bucket, and the time it took for the bucket to be filled in order to calculate how much of the bucket each person had filled—it might get complicated, but in principle one could separate out the independent contributions of genes and environment on trait development. But in the world of contingency and co-constitution, one person would be in charge of controlling the faucet and the other would be responsible for positioning the hose above the bucket—the consequences of each person's actions would be inextricable from the actions of the other, and they could only fill the bucket together. And one might drop the hose or slip while turning the faucet handle, or a strong gust of wind might spray some of the water outside the bucket—the time taken to fill the bucket depends on contingent events too. The very question of how much "either" genes "or" environment contribute to a trait becomes, in this contingent and co-constituted world, entirely meaningless—they're simply impossible to separate.

MELINA AND PEPPER

For our second example of how animal behavior emerges from the co-constitution of genetic, environmental, and social forces, we turn to Pepper, the dog who made Melina fundamentally question her feminist skepticism of biology and led her to a more nuanced understanding of the "genes versus environment" debate.

Feminists (among others working toward social justice) have good reasons to reject biological determinism. Biological determinism lends support to the patriarchal idea that females, including human women, are *inherently* fill-in-the-blank (emotional, nurturing, coy, passive, and what not), irrespective of where, when, or how they were socialized. Human and animal males are similarly understood as *inherently* fill-in-the-blank (aggressive, rational, better at math and science, profligate, and so on), thus justifying patriarchal social relations. Whereas Bateman tried to argue that stereotypically gendered behaviors were simply a result of adaptation by natural selection in males and females (see chapter 2), biological determinism locks these stereotypes into an animal's biology, specifically their genes, no matter the animal's social and environmental conditions. Refusing these stereotypes, feminists firmly reject the belief that there is some sort of biological, genetic, immutable "essence" to being female or male, whether referring to humans or animals. But at some point in her feminist training, Melina overcorrected, insisting that genes have practically *nothing* to do with behavior. She believed firmly in nurture over nature.

Then she adopted a dog. A "purebred" dog at that, who was bred and trained to hunt birds: an English Pointer. After years of intently observing and learning alongside Pepper, and, initially, making lots of rookie mistakes (many at his expense, unfortunately), Melina simply could not deny the material, visceral effects of centuries of dog breeding on his hunting and pointing behaviors. (These dogs "point" by freezing in front of the prey animal, usually a bird, and gesturing toward the prey animal's direction for the benefit of the hunter, indicating with their nose, paw, and/or tail; see figure 7.2.) For instance, Melina thought Pepper was being disobedient—and that she was a bad dog handler—when he sprinted well ahead of her and seemed to completely ignore her when she called him back. After working with dog trainers who specialize in hunting breeds, and reading several Pointer training books on her own, Melina came to understand that Pepper basically cannot hear her when he's so excited and distracted by whatever prey animal he's running toward. She also learned that a dog like Pepper is *supposed* to run hundreds of yards ahead—he was probably expecting her to hunt behind him—so he was actually being a *good* dog all those times she thought he was naughtily running away.

Witnessing Pepper point pushed Melina to allow for the possibility that there are indeed significant biological components to his behaviors. Many animals, including humans, point naturally—pointing, in this context, is the pause before the pounce, just as you might steady your hand above a fly you're about to swat. When Pepper points, he is basically freezing mid-step, just before pouncing on the prey animal. What makes Pointers different from other dogs is their ability to *hold* their point, or continue to stay frozen even as the bird they've located starts to fly away, so that the hunter can take

Figure 7.2
Pepper on point. Photo: Melina Packer.

a shot at the bird. It's nearly impossible to get Pepper off a point without physically dragging him, and most of the time he's far out of Melina's sight anyway, which makes pulling him away from a point a non-option.

Even after many generations of selective breeding, most Pointers still have to be trained to hold their point. Pointers were thus selected by hunting dog breeders not just for their pointing tendencies but also for their trainability—some individual dogs learned to hold their points better and faster than others, and only those dogs who happened to perform these human-desired behaviors more successfully were continuously bred. (Clearly the process of using artificial selection to breed all sorts of domesticated animals, including livestock, pets, and "working dogs" like Pepper, is yet another application of eugenics.)[21] But while Pepper appears to naturally enjoy running and pointing, and moreover seems naturally unable to stop running and pointing, Melina knew she could not characterize his desire to run and point as simply biological, given centuries of selective breeding and generations of specialized training. As real as Pepper's genetically informed behaviors are, such biological happenings would not be happening without the socioeconomic structures supporting them, including the leisure time and expansive properties to hunt with uniquely bred and trained "bird dogs," not to mention the ancient (but not universal) belief that Man may and ought to manipulate Animal for his own benefit. To attribute Pepper's "natural" behavior to his biology alone would be to ignore the historical and ongoing roles of human desires and interventions in shaping his genetics.

Breeding and training dogs to hunt birds and yet not hunt birds—to persist obsessively in finding birds but ultimately to resist the urge to kill them (saving that thrill for the human hunter)—is quite a contradiction. And arguably, breeders have bred hunting dogs into a corner: pointing dogs like Pepper have been bred and trained, over centuries of artificial selection, to have an extremely intense prey drive, which then must be contained and controlled in order for the human hunter to finish the job. How do hunting dog handlers produce this whiplash-like behavior? With a tremendous amount of physical force. Over the past four hundred years (at least), hunting dog handlers have used all manner of tools to force hunting dogs *not* to kill or eat the birds they have been trained to find and point, or in some cases retrieve.[22] Ropes, nets, sticks, chains, electric collars around the neck, electric collars around the flank, beatings by hand—you name it. Given such forceful training methods, hunting dog breeders have also been motivated to artificially select for pliant and obedient dogs. As one of Darwin's students observed: "The whole character of the dog may . . . be said to have been moulded by human agency with reference to human requirements."[23] Pointers today are exceedingly gentle (on average), tolerating a tremendous amount of physical force against them while remaining nonaggressive.

Of course, not all hunters use force to train their dogs—quite the opposite in fact. A quick glance through any contemporary hunting magazine or social media account will attest to the fact that human hunters overwhelmingly love and even pamper their hunting dogs, considering them close companions and members of the family. And anyone who knows Melina knows how much she loves and dotes on—some might say obsesses over—Pepper. Yet she continues to rely on an electric collar in order to effectively handle Pepper, after receiving extensive training on how to use this tool humanely. Melina has had to humbly and radically accept that the biological and social contingencies of history have produced a dog who cannot enjoy a pleasurable life without an electric collar, as Pepper is simply impossible to handle (for his own safety and well-being) without one, through no fault of his own. And despite all of Melina's efforts at giving Pepper the physical and mental and emotional outlets he has been bred and trained to need—huge fields to run through off-leash, countless birds to find and point, all while using the least amount of force possible—he is an extremely difficult dog to keep satisfied and happy in today's overwhelmingly urban, industrialized world. It's almost as if Pepper belongs in a different time and place, where land is endless, motorized vehicles do not exist, and nobody needs to get any work done. (A time and place that was perhaps always an exclusive privilege of the rich, landed gentry of Britain.)

So, to sum up, pointing dogs like Pepper are all-too-real products of *both* genes *and* environments, simultaneously and inextricably. Pepper's so-called natural, biological behaviors would not exist, and cannot be understood apart from, their historical, social, and environmental contexts. This does not mean that biology and genetics are mere social constructions with no material consequences. On the contrary, dogs like Pepper may be understood as *victims* of artificial selection, given that the behavioral traits that were bred into them can be a serious liability in today's world. As Sunaura Taylor put it (referring to other kinds of pets and livestock): "These animals are simultaneously disabled and hyperabled—made disabled by the very enhancements that make them especially profitable to industries and desirable to consumers."[24] Which is to say, the social construction of biology does not make it any less real. All the more reason to take much greater care in constructing the cultural contexts that shape how biological processes may unfold.

TROUBLING THE BINARY

Biological determinism tells us that an individual's biology, specifically their genes, determines whether they possess eggs or sperm, which in turn determines what the individual will look like and how they will behave.[25] Indeed, within the paradigm of biological determinism, patriarchal stereotypes of female and male behavior—the coy females and profligate males we first encountered in chapter 2—can be traced directly to the differences between sperm and egg. Eggs are larger and less numerous than sperm, and differences in male and female sexual behavior are said to follow directly from these differences in size and number. Because there are fewer eggs than sperm, the logic goes, and fertilization requires one sperm and one egg, there must be greater competition among males for opportunities to fertilize eggs and a desire in males to fertilize as many eggs as possible. By contrast, the logic continues, females must take great care not to waste their limited, resource-rich eggs by allowing them to be fertilized by sperm from subpar males.[26] Ergo, eager males and choosy females.

Over the last four decades, feminist science studies scholars have refused determinism over and over again, resisting the idea that females are "naturally" some combination of choosy, coy, conniving, maternal, and domestic—a suite of stereotypes that infiltrate how animal behavior scientists think about sex as well as gender. In her enduringly popular 1991 essay "The Egg and the Sperm," cultural anthropologist Emily Martin closely analyzed major US biology textbooks to demonstrate how the egg and the sperm are characterized as stereotypically feminine and masculine, respectively. Textbook authors tended to describe egg cells as passive, wasteful, and disposable, while sperm cells were strong, brave, persistent, and productive. Moreover, when researchers and journalists wrote about newer scientific findings that showed egg cells to be more active participants in fertilization than previously thought, and sperm cells to be less forceful, authors continued to fall into gender stereotypes: instead of passive and maternal, eggs were now described as manipulative. As Martin summarized: "New data did not lead scientists to eliminate gender stereotypes in their descriptions of egg and sperm. Instead scientists simply began to describe egg and sperm in different, but no less damaging, terms."[27]

Why does it matter how scientists and journalists describe cellular processes involving single gametes? It matters, Martin argued, because such "imagery keeps alive some of the hoariest old stereotypes about weak damsels in distress and their strong male rescuers. That these stereotypes are now being written in at the level of the *cell* constitutes a powerful move to make them seem so natural as to be beyond alteration."[28]

This binary sex categorization of gametes also extends to gonads, hormones, and the entire animal body, thereby reinforcing the role of biological determinism in the perpetuation of binary sex categories. For example, historian of science Nelly Oudshoorn's 1994 book *Beyond the Natural Body* documented how the scientists who discovered hormones in the early twentieth century layered their cultural norms about human gender roles onto this newfound biological material.[29] Early endocrinologists noted, for instance, that both females and males possess hormones such as estrogen and testosterone, but nevertheless decided to categorize estrogen as a "female hormone" and testosterone as a "male hormone." These same scientists found that bodies use hormones like estrogen and testosterone for all sorts of developmental and metabolic processes that are unrelated to sex or reproduction, and yet they named these hormones "sex hormones." In doing so, these scientists had foregone an opportunity for a radical redefinition of sex categories. They could have interpreted their data on the wide ranging roles of estrogen and testosterone in animal bodies to suggest that "chemically speaking, all organisms are both male and female."[30] But, again, new data did not lead scientists to eliminate gender stereotypes. Endocrinologists continued to use sexed terminology, promoting the idea that hormones determine sex (and sexuality) and moreover that testosterone generates masculinity while estrogen generates femininity. Feminist science studies scholar Sarah Richardson documented a similar fate of the X chromosome, which has come to represent the genetic source of femininity, even though female sexual development is shaped by a multitude of genes across many chromosomes.[31]

A feminist approach to scientific data and narratives shows us how messy sex categorization is for all kinds of biological traits: behaviors, hormones, gonads, chromosomes, and gametes. Moreover, these different indicators of sex, each messy on their own, do not align with one another to form neat columns into which we can divide animal bodies.[32] As biomedical ethicist Kristina Karkazis and her feminist science studies colleagues explained in 2012: "The breadth of human physical variance is more complex than the categories [of female and male] suggest."[33] They offered the example of complete androgen insensitivity syndrome (CAIS), a diagnosis applied to women

> who are born with XY chromosomes, testes, and testosterone levels in the typical range for males. If only taking chromosomal, gonadal, or hormonal factors into account, one would label these individuals male. Yet these women have a completely feminine [external] phenotype, with breast development and female typical genitalia, because their androgen receptors are not responsive to androgens. Designating women with CAIS as male would be inappropriate, given that they are presumed female at birth, are raised as girls, and overwhelmingly identify as female. . . . In the context of reproduction, . . . [a] woman who has undergone a hysterectomy has no uterus in the same way a woman with CAIS has no uterus, yet no one questions whether the former is really still female.[34]

But such sex-related complexities are not unique to people with such "disorders." *None* of the physiological attributes that scientists designate as sex markers—gonads, gametes, hormones, chromosomes, and so on—are binary. Each sex marker "contains significant variation, both within and across individuals. For example, women's testosterone levels range widely among women and also by time of day, time of month, and time of life."[35] Ultimately, Karkazis and colleagues reminded us: "there is no single physiological or biological marker that allows for the simple categorization of people as male or female."[36]

Much the same is true of animals. In chapter 1, we learned about American green frogs who switch sexual expression during the course of their lifetime, possibly due to temperature variation, resulting not only in adult frogs who are chromosomally "male" yet have ovaries, but also in adult frogs with egg-producing cells in their testes, for example.[37] Or consider variation in testosterone, which, across animals, is generally associated with gametes —sperm-producing individuals tend to have more testosterone than egg-producing individuals. But variation in testosterone also depends upon individuals' social interactions and roles, and this fact is as true for humans as it is for northern jacanas, yellow-legged birds that live in the wetlands of Central America.[38] Many northern jacanas who produce sperm also construct rudimentary nests and incubate eggs during the breeding season, and have relatively lower levels of testosterone while doing so. Mainstream narratives in animal behavior science try to force this variation into the boxes of binary sex, first categorizing individuals as male or female by assuming that sex is determined by gametes and then describing these birds as "sex-role reversed" because the jacanas categorized as males have more "feminine" social behaviors than those categorized as female. In this and many other animal species—hummingbirds, damselflies, seahorses, white-tailed deer, dolphins, whales, and hyenas, to name just a few—forcing our observations of sex-linked traits into a binary confuses rather than clarifies.[39] When we free ourselves from the expectations of determinism, this kind of forced binary reveals itself to be nothing more than an ideological commitment with limited scientific utility. By insisting upon a singular way that animals *should* look and behave, binary sex categories prevent us from appreciating and understanding how animals *actually* look and behave. The insistence on binary sex categories happens *despite* biological reality, not because of it.

The late nineteenth- and early twentieth-century scientists who created binary sex categories themselves grappled with their categories' incoherence. As historian and feminist science studies scholar Beans Velocci wrote, describing these early years of sex science:

> [There] was a vast quantity of evidence that did not at all show an obvious or stable division between male and female bodies, or a singular thing called sex at all. If anything, as scientists dug ever more deeply into sex, they created more . . . problems than they solved. Non-human species displayed an enormous range of possibilities for how bodies could be arranged and how they might reproduce: female hyenas with what looked to be phalluses, bees with three sexes, startlingly frequent incidences of cattle with ambiguous sexual characteristics. Attempting to pin down what made someone or something male or female dredged up endless counterexamples as scientists debated whether sex was a matter of gonads, metabolic rate, chromosomes, or something else entirely.[40]

What these scientists, and many scientists after them, have ignored or denied is that the female/male sex binary is far more a product of culture than of biology; after all, different cultures in different historical periods have had different yet entirely meaningful categories of sex.[41] Moreover, as cultural theorist Judith Butler argued, it is the existence of gender norms that lead scientists (among others) to care so much about defining sex categories in the first place, meaning that sex itself, no matter how material, is always already cultural and political, and these political stakes can extend beyond patriarchy to other systems of oppression as well.[42]

The scientific project of assigning sex categories to organisms was intimately related to white supremacism. In *The Descent of Man*, for example, Darwin concerned himself with traits in humans that he thought were more pronounced in women than in men, and believed that these traits are also "characteristic of the lower races, and therefore of a past and lower state of civilization."[43] European and Euro-American categories of race, sex, and animality thus must be understood in relation to one another—they were co-constituted.[44] As discussed in chapter 6, European and Euro-American scientists classified people of African descent as the lowest on a hierarchical scale of human races. They justified this ranking in part *because* they perceived African men and women to be less distinguishable from one another, or less sexually dimorphic (as opposed to men and women of the "white race"). Racial science thus declared that more evolutionarily advanced humans were also more sexually dimorphic, and "the sex/gender difference that is supposedly displayed fully only by the European heterosexual couple serves as an ideal against which to measure all races."[45] Dominant assumptions about "normal" female or male anatomy, or "advanced" feminine or masculine behavior, are thus based upon a historically specific white, Christian, heterosexual human standard, which is assumed to be superior to a more "animal-like," racialized Other.

Moreover, in the US, much of the work of scientific sex categorization was tackled by committed eugenicists, employed at research institutions like the Station for Experimental Evolution and the Eugenics Record Office.[46] These scientists' work of classifying sex was inextricable from their work of controlling who gets to reproduce and who does not, with the goal of maintaining the dominance of the white, the heterosexual, the able-bodied, and the wealthy. What's at stake with binary sex categorization, therefore, is the power to control the lives of those seen as less worthy. And by positioning themselves as the experts on sex categorization, animal sex scientists held tightly onto power.

When viewed through the lens of power, scientists' insistence on binary sex categorization, despite its empirical incoherence, makes sense. If determining sex categories is both complicated *and* essential to political power, then those experts who can distinguish the sexes can make a particular claim to power. (More generally, if behavior is determined by biology, then we *need* biologists in order to understand, and control, behavior.) As Velocci observed:

> For [the] eugenicists, there was no unified thing called sex. Rather, sex functioned as an umbrella category for a constellation of things that were and remain assumed to clump together in sometimes contradictory ways: the body, the mind, reproduction, behavior, hormones, gametes, morality, family role, a square or circle on a form. *What mattered was what would best serve the eugenic mission.* There is not, for scientists in the past or historians [in] the present, a pre-existing thing called sex whose boundaries can be precisely mapped. Only something made by a set of research practices, responding, in the case of eugenics, to a desire for white domination.[47]

Rigid and deterministic definitions of sex continue to be used to bolster oppressive policies and laws, lending these policies and laws the false authority of science. As we drafted this chapter, for instance, we learned that Canada has issued a travel warning to LGBTQIA+ folks for the United States. Why? Because numerous US states have recently passed legislation banning teachers from talking about sexual diversity in schools and libraries, banning trans people from public restrooms and from participation in sports, and criminalizing gender-affirming medical care, among other civil rights and free speech violations.[48] Such legislation necessarily rests on binary and deterministic conceptions of sex; see, for example, the definition of sex employed in the 2023 Arizona State Bill 1702, which sought to ban gender-affirming health care: "'Biological sex' means the biological indication of male and female in the context of reproductive potential or capacity, such as sex chromosomes, naturally occurring sex hormones, gonads and nonambiguous internal and external genitalia present at birth, without regard to an individual's psychological, chosen or subjective experience of gender."[49]

In order to advance their discriminatory, transphobic agenda (to say nothing of their erasure of intersex people), the authors of this bill must claim—inaccurately—that binary sex categorization is possible. It is imperative that biologists publicly refute and condemn such false appeals to scientific objectivity.[50] More generally, it is essential for biologists to become a part of the necessary shift in how sex is researched and reported in scientific scholarship and mainstream media alike. For queer, trans, and nonbinary people, how scientists describe sex is truly a matter of life and death.

When the biological sciences uphold the dominant cultural myth of binary sex categories despite the evidence, then all the social, political, medical, and economic injustices that queer and gender-nonconforming people face would seem to be naturally, biologically justified. To enact justice in opposition to biology is not impossible, but becomes a struggle *against* nature. In contrast, happily accepting that sex is an impossibly messy category opens up new possibilities for our understanding of animal biology and behavior, as well as for our understanding, and treatment, of ourselves.

What, then, is a more accurate, evidence-based scientific view of animal sex that allows us to reject the bigoted politics of control over sexual expression (supported by binary sex categorization) and instead embrace an expansive and liberatory politics? The answer begins with rejecting determinism and instead centering the messy variability of the multitude of traits that we currently pile into the category of sex. As trans biologists and neuroscientists Miriam Miyagi, Eartha Mae Guthman, and Simón(e) Dow-Kuang Sun wrote in a 2022 essay: "Rather than privilege any characteristic as the sole determinant of sex, 'male' and 'female' should be treated as context-dependent categories with flexible associations to multiple variables . . . No one trait determines whether a person [or animal] is male or female, and no person's [or animal's] sex can be meaningfully prescribed by any single variable."[51] This liberatory definition of sex depends upon eschewing determinism and embracing contingency. When we understand that an animal's life isn't predetermined but rather unfolds contingently in a specific context, we immediately make room for variation in sexual expression.[52] Traits that we consider relevant to sex expression are not *necessarily* associated but rather *tend to be* associated, to greater or lesser degrees, because the multitude of causal processes that shape an animal's sexual development are contingent upon an astounding array of factors and processes, including interactions between these factors and processes—another instance of co-constitution.

Certainly, the many different variables that we place into the fuzzy category of sex are loosely correlated—egg-bearing humans often also have more breast tissue, for example. But when we shed the allegiance to strict correlations among these variables, we release the expectation of a sex binary. And once we allow for loose associations and plenty of variation within this sprawling set of characteristics that make up "sex," we must contend with exactly what we mean by "sex" *every time* we use the word or concept, and our deliberation must include a consideration of the political stakes. As philosopher Hane Htut Maung described, in a 2023 paper, "the boundaries of femaleness and maleness are not simply given, but have to be negotiated. Such negotiation is not only informed by the empirical data, but also by our values and interests relative to various contexts and purposes."[53] The more we reject the simplistic conclusions of biological determinism and instead engage in these nuanced, context-specific deliberations about what exactly we mean when we say "sex," the more we will come to question whether the fuzzy category of sex is useful at all, scientifically, socially, or politically.[54]

What becomes possible when we stop trying to fit animals into binary sex categories? In a more queer and trans conception of animal sex, animals would not be anomalous or paradoxical because of their sexual expression—they would simply be. American green frogs and northern jacanas would be interesting to us not because they fail to conform to our expectations, but because, just like any other animal, they embody a unique way of persisting through time. And as we've seen throughout *Feminism in the Wild*, it's not just the expectations of binary sex categories that animals are trapped within—they are trapped within the constraining expectations of *multiple* systems of power. Thus, every animal we encounter—Coho salmon, anole lizards, lance-tailed manakins, rabbits, dogs, dung beetles, ravens, hens, caterpillars and fruit flies—has countless stories, anchored in diverse, liberatory standpoints of all kinds, that we can tell about them. It's this expansive collection of stories that will move animal behavior science closer to justice, for humans and animals alike.

DOING SCIENCE DIFFERENTLY

We authors would be remiss if we didn't spend a moment talking about feminist scientific methods, which, in the case of animal behavior science, requires engagement with feminist animal ethics. If we, as feminist observers of animals, whether for research or for pleasure (or both!), are intentional about refusing hierarchies of power, including hierarchies of cognition, then how can we justify conducting experiments on animals? Keep in mind that most of these experiments require capturing individuals from the wild and euthanizing them at the end of the study (to say nothing of the millions of organisms bred solely for scientific research).[55] Arguably, even an animal who's been captured for the purposes of a scientific experiment and then released relatively unharmed back into the wild has had their autonomy violated by human hands, however well-meaning those humans might be.

And yet much of what we humans know about animals today, domesticated and wild, comes from some form of human experimentation on animals, whether through centuries of artificial selection or months of laboratory studies. Even research designed to demonstrate animal agency and intelligence, including empathy, often relies upon some painful, traumatizing, or at least highly inconvenient and nonconsensual intrusion on animals' lives.[56] So the question remains: how can we study and produce scientific knowledge about animals, and otherwise more equitably share the world with animals, in ways that acknowledge and respect animal agency, and attend to every animal's unique senses of pain and pleasure?

Donna Haraway wrote at length on animals "suffering for science" in her 2008 book *When Species Meet*.[57] She argued that researchers must "do the *work* of paying attention and making sure that the animal suffering is minimal, necessary, and consequential. If any of those assurances are found impossible . . . then the responsible work is to bring the enterprise to a halt."[58] But a particular challenge of practicing feminist animal behavior science is that, however much we might want to, it remains difficult for us humans to fully understand animals' needs or experiences. Feminist animal ethicist Lori Gruen offered the approach of "entangled empathy" in her 2015 book of the same title. Rather than hold our human selves at a distance from (and largely above) animals, Gruen urged us to accept and grapple with our entanglements with animal others, embracing the reality that our human emotions, and politics, are intimately tied up with our scientific rationales, and that animals have situated knowledges and standpoints of their own. Gruen's feminist animal ethic of entangled empathy thus invites us "to attend to other animals in all of their difference, including differences in power within systems of human dominance in which other animals are seen and used as resources or tools."[59]

As we continue to turn away from categories and hierarchies that divide humans from animals, and that divide different humans from one another, we may find that a truly feminist science of animal behavior will reject animal experimentation altogether, opting instead for minimally intrusive observations of animals in the habitats in which they already live. Nevertheless, we authors firmly believe that refusing mainstream science's tacit Man-over-Animal hierarchy does not mean curtailing our knowledge. Rather, letting go of hierarchies (and singular stories) opens up vast new worlds of possibilities, empowering us to co-create even greater knowledge about animals, and about ourselves. Relinquishing hierarchy allows us to build, together, an expansive universe that values the intrinsic worth and unique *Umwelten* of each and every entangled being: animal, human, and otherwise.

In making this transition away from mainstream science and toward a more feminist approach, it can be difficult to know where to begin, to find the questions about animal behavior that are worth asking. We leave you with an exhortation, from Levins and Lewontin, that we hold as something of a north star. They suggested letting go of the belief, seemingly central to mainstream science, that a scientific question is meaningful if it "is logically well defined, testable, and capable of being answered on its own terms without regard to application." Instead, they argued: "a question is meaningful if what we *do* or *feel* is changed by the answer."[60]

We hope that *Feminism in the Wild* inspires you, dear reader, to *feel* differently about animals, and, for the scientists among us, to therefore *do* different science, to tell different stories about what our world is and what it can be, to see nature, and ourselves, in a more kaleidoscopic and more wondrous light.

Acknowledgments

AMBIKA: THE PEOPLE AND OTHER ANIMALS WHO BROUGHT ME HERE

To the antlions, lizards, ravens, and tent caterpillars—I don't have words for how much you've made me.

I cannot imagine a more generous, conscientious, patient, principled, knowledgeable, and astute collaborator than Melina. I am a sharper thinker and better feminist because of her, a gift that extends through the pages of this book and far beyond. I'm so lucky she agreed to work on this project with me!

Tremendous gratitude to my mentors in behavioral ecology and evolutionary biology: Suhel Quader, Damian Elias, and Jonathan Losos. I was lucky to learn how to be a field biologist from Yoel Stuart, the best scientist I know. Thank you to all my labmates over the years, especially Martha Muñoz, Alexis Harrison, Sofia Prado-Irwin, Nick Herrmann, Shane Campbell Staton, Talia Yuki Moore, Pavitra Muralidhar, Thom Sanger, Melissa Kemp, Ian Wang, Anthony Geneva, Oriol Lapiedra, Claire Dufour, Kristin Winchell, Colin Donihue, Ignacio Escalante Meza, Erin Brandt, Maggie Raboin, Trinity Walls, and Malcolm Rosenthal, as well as my amazing field assistants: Chetan Kokatnur, Divyaraj Shah, Rachel Moon, Christian Perez, and Jon Suh. Much appreciation for the Organization for Tropical Studies (South Africa fall 2009 and Costa Rica winter 2012), where I learned to love natural history. And a huge thank you to the Miller-Levin Lab at Amherst College, for being much-needed botanical respite from the world of animal behavior.

Zuleyma Tang-Martínez, thank you for being the elder our field has needed—you are the giant on whose shoulders my work rests. For your encouragement and wisdom, at various times: Marlene Zuk, Doug Emlen, Dan Bolnick, Jeremy Fox, Meg Duffy, Andy Warren, James Costa, Armin Moczek, Mike Wade, Kim Hoke, Banu Subramaniam, and Ben de Bivort. Thanks to Norm Ellstrand for writing the funniest, most serious paper ever, and emailing with Yoel and me about it—that exchange was one of many seeds that grew into this book!

My behavioral ecologist and evolutionary biologist colleagues, thank you for thinking together, for all the advice, shared struggle, and conference camaraderie: Maria Rebolleda Gomez, Vince Formica, Nancy Chen, Julia Monk, Caitlin McDonough, Max Lambert, Erin Giglio, Liam Taylor, Harold Eyster, Ben Allen, Sreekar Rachakonda, James Crall, Christie Riehl, Cassie Stoddard, Michele Johnson, Swanne Gordon, Courtney Fitzpatrick, Amanda Hund, Helen McCreery, Kiyoko Gotanda, Kinsey Brock, Alison Feder, Brandon Ogbunu, Cathy Rushworth, Katie Ferris, Alex Gunderson, Hannah Frank, Scott Taylor, Sara Lipshutz, and so many others.

Liz Neeley, thank you for validating my writerly ambition! Liz, Victoria Fine, Sue Pierre, Chelsie Romulo, Simon Donner, and everyone else at the inaugural 2022 StoryMakers workshop—you're wonderful. Ed Yong, thanks for answering my slightly panicked texts about how to actually write a book. Thanks to ComSciCon, and my journalist friends and colleagues (especially Nidhi Subbaraman, Darren Incorvaia, and Asher Elbein) for the advice and articles. Nandini Nair, Soity Bannerjee, and Veena Venugopal—abundantly grateful for the chance you gave me to write about animals delightedly, over and over. Conversations with Tisse Takagi were invaluable early in the process of figuring out what this book needed to be.

Whitney Robinson, thank you for the space to salvage so many of my Fridays with writing! Lisa Haney, Mark Friedman, and Muse the cat, thank you for lending me your home—the perfect writing context and company.

Michiko Theurer and Sama Ahmed, in very different ways, the two of you are my absolute favorite people to think and feel and dream with. It's been such a joy to work on this book in the company of our friendships. Dan Fourie, thank you for helping me stay connected to wonder, audacity, and honesty at the early stages of this project, and for introducing me to systems thinking and affordances. David Jay, our very intense walk-and-talks were so much fuel—beyond glad that we met when we did! Ned Burnell, thank you for organizing me into seeing myself as an organizer, for agreeing all those years ago that I had a book in me, and for continuing to believe.

All my favorite people, my buoys in some choppy waters (in no particular order): Grace, Holly, Mara, Jasmine, Dodi, Vinski, Ashton, Beans, Kayleigh, Ned, Dan, Garret, Joey, Mark, Yoel, Max L., Max G., Nick, Sofia, Michi T., Lauren, Scott, Jeremiah, Colette, Joshua, Madeleine, Jason, Meg, Kit, Kathy, Leann, Marcy, Teresa, Mary Ann, Courtenay, Amanda, Benji, Didem, Nidhi, Irina, Dorian, Esteban, Yong, Alex G., Ransom, David, Julie, Sandra, Sophie, Michiko M.W., Alex H., and everyone at UAW5810 and TCC—love you all, thank you for keeping me here and happy. Ansel, you are exactly my intended audience! I love how much our conversations and our relationship have shaped this book, and me, and I love you.

Ayodh Kamath, you made me a scientist, and that saved me. Somehow, you know so much of my mind, and I wouldn't have it any other way; thank you for everything. Jayanthi Subramanyam, thank you for being the rock we needed.

MELINA: THE PEOPLE AND OTHER ANIMALS WHO BROUGHT ME HERE

I wish I knew whether Pepper, my canine companion, likewise knew how magnificently and profoundly he has shifted my thinking, feeling, and being. Something in his haughty yet entirely endearing air of entitlement tells me that he does. I love how unapologetically and assuredly he moves through the world, ever-present in his body and a constant, generous reminder that there is no mind-body divide. So whether he knows it (or cares) or not, I am deeply grateful to Pepper for teaching me to vastly expand my sense of how to be human, and how to be in better relation with other beings, human and more than human.

I am just as grateful to Ambika for not only teaching me so much about science, but also for being living proof that a truly feminist, anti-racist, anti-ableist, queer, and Marxist science is possible! She is a fantastically fierce person to write and think with in general, and I am honored to have contributed to such an important political and personal project.

Huge thanks are also due to my ecotoxicologist interlocutors: I can't name any names but you know who you are.

My intellectual crushes are too many to list, but here I would like to highlight the scholarship and mentorship of Traci Brynne-Voyles, Jake Kosek, Nicole C. Nelson, Banu Subramaniam, Charis Thompson, Harlan Weaver, and Cleo Wölfle Hazard. Your work is foundational to my own, and, more importantly, you've labored to put it out into the world for people to learn from and behave better, as a result. Plants and animals urgently need more humans who think and act like you do; thank you.

How fortunate I am to now have colleagues at the Universities of Wisconsin who are as excited about my queer feminist and anti-racist animal studies research as I am, particularly new friends in the UW La Crosse Departments of Race, Gender, and Sexuality Studies, English, and Environmental Studies.

Reading and thinking and writing and acting of course never happen in a vacuum; thanks especially to Brett C., Vera C., Ashton W., Tara G., Laura D., Julie G., Annie L., Julie W., and Scott J. for enriching my life with compelling conversation, good humor, adorable babies, refreshing exercise, fabulous music, delicious food, and intoxicating substances, in no particular order. I am principally indebted to B. N. for many an intellectual (and emotional) breakthrough. I hope this book helps you liberate yourself from your Hobbesian tendencies, B., or at least to see them as a choice, rather than an inescapable fate.

Where would Pepper and I be without our "dog men" (who will be thanked more profusely in my dog book)? Alec, George, and Arman: even though Pepper and I didn't always cooperate, we learned so much from you with every training session, every shooting practice, every equipment lesson, and every feisty debate. I hope that each of you learn something from this book and, in turn, shift some of your positions about what dogs "naturally" do (or don't do).

Winona, Zack, and Lang: thanks for all the hilarious animal memes and Pepper-related triage. Mom: I hope this book pushes you to reconsider your disdain for certain mammals (who will remain nameless); thank you for reading all the same. Dad: nothing is more animal than my grief in your absence.

And finally, Shane: your love of (my love of) animals is one of innumerable reasons why I love you. Pepper, Rex, Iman, and I are so much happier with you in our multispecies cozy. Thank you.

THE PEOPLE WHO, AND INSTITUTIONS THAT, MADE THIS BOOK POSSIBLE

To our acquisitions editor, Anne-Marie Bono, for believing in our idea and making this book a reality! Our gratitude and appreciation also extends to the proposal and manuscript reviewers whose thoughtful feedback improved this project substantially. Huge thank you to everyone at the MIT Press who worked on transforming this book from a manuscript to its final form: Caroline Helms (acquisitions assistant), Ginny Crossman (editor), Natalie Jones (copy editor), Katie Kerr (art project manager), Emily Gutheinz (senior designer), Kate Elwell (production coordinator), Debora Kuan (copywriter), and Helen Weldon (publicist).

To the hundred-odd students in EBIO3240 in the fall of 2022 at the University of Colorado Boulder—thank you for taking our strange course into your stride, and maybe even getting excited about it. Our colleague Caitlin Kelly was crucial to getting this experiment off the ground, and our teaching assistants—Katie Mazalova, Alec Chiono, Gladiana Spitz, and Aaron Westmoreland—helped us pull it off. Further thanks to Aaron and the dozen or so students in EBIO6200 Spring 2023 for thinking through some controversial ideas in evolutionary theory with Ambika. A huge thanks to Lucia Frias-Wackman, Biman Xie, and Hannah Gellert: your brilliant and honest feedback on our course materials made this a much better book!

To the Feminist Lenses for Animal Interactions Research (FLAIR) lab. Katie Mazalova, Jenna Pruett, Dylan Clements, Yanru Wei, Sam Rothberg,

Maggie Silver—your presence, acumen, and enthusiasm made our work together so joyful.

The creation of this book was supported, financially and intellectually, by the Department of Ecology and Evolutionary Biology at the University of Colorado Boulder. Thank you for giving our unusual work a welcoming home, and for all the resources we needed to bring this book into existence. The Miller Institute for Basic Research in Science at the University of California Berkeley funded Ambika in the early stages of this project.

Our indispensable intellectual comrades include Ashton Wesner, Max Lambert, and Beans Velocci. We all met through the *Queer Ecologies | Feminist Biologies* working group that Ashton brilliantly assembled in 2018, funded by the UC Berkeley Social Science Matrix. All three of you deepen our understanding of what science is, and help us imagine what science can be; we are so excited to continue thinking and writing with you. Thanks especially to Beans Velocci and also Gabe Winant, for replying to Ambika's *many* texts about history in general and transness, eugenics, dialectics, and Marxism in particular. Jessica Riskin, thank you for writing *The Restless Clock* and for chatting with Ambika about it. Emily DuVal, Sara Lipshutz, Max Lambert, and Jonathan Losos, thanks for sharing your deep expertise on manakins, jacanas, American green frogs, and anoles, respectively, with us!

Mary Decker and Edy Scripps offered early feedback and encouragement. Bill Bank, thank you for copyediting the whole first draft! Ansel Schmidt, Alec Chiono, Ignacio Escalante Meza, Sam Rothberg, Michiko Theurer, Sama Ahmed, Ned Burnell, Henry Richardson, Liam Taylor, Kim Hoke, Jenna Pruett, Katie Mazalova, Ashton Wesner, Max Lambert, Beans Velocci, Yoel Stuart, Anne Walsh, Valerie Young, and Lauren Chun read all or part of the manuscript, and gave us incredibly helpful critique and suggestions while cheering us on most generously—we're so grateful! And last, thank *you*, dear reader, for joining us on this journey.

Notes

CHAPTER 1

1. Charles Darwin, *On the Origin of Species by Means of Natural Selection* (London: J. Murray, 1859).

2. Jason V. Watters, "Can the Alternative Male Tactics 'Fighter' and 'Sneaker' Be Considered 'Coercer' and 'Cooperator' in Coho Salmon?," *Animal Behaviour* 70, no. 5 (2005): 1055–1062, https://doi.org/10.1016/j.anbehav.2005.01.025.

3. The hooknoses are not always successful at chasing away their competition; in Watters's observations, "on average females were attended at any one time by three potential mates," including jacks and hooknoses. Watters, "Can the Alternative Male Tactics 'Fighter' and 'Sneaker' Be Considered 'Coercer' and 'Cooperator' in Coho Salmon?," 1057.

4. Erika King et al., "Alternative Life-History Strategy Contributions to Effective Population Size in a Naturally Spawning Salmon Population," *Evolutionary Applications* 16, no. 8 (2023): 1472–1482, https://doi.org/10.1111/eva.13580.

5. King et al., "Alternative Life-History Strategy Contributions."

6. To make absolutely sure of this claim, we'd have to do an exhaustive search of the scientific literature on Coho salmon mating, which would be a gargantuan task—a search on Google Scholar in October 2023 for "'coho salmon' + mating" yields over twenty-three thousand results. By examining citations of Watters's 2005 study, however, we can be reasonably certain that his hypotheses on Coho salmon sex have not been tested further. Watters himself no longer does research on Coho salmon behavior, and instead has spent his career in animal welfare at the San Francisco Zoo.

7. It is no coincidence that Victorian England was a time of considerable shifts in gender roles, perhaps motivating mostly male researchers to cement traditional gender roles in "nature." See, for example, Cynthia E. Russett, *Sexual Science: The Victorian Construction of Womanhood* (Cambridge, MA: Harvard University Press, 1995).

8. Sarah Blaffer Hrdy, "Empathy, Polyandry, and the Myth of the Coy Female," in *Feminist Approaches to Science*, ed. Ruth Bleier (New York: Pergamon Press, 1986), 119–146; Zuleyma

Tang-Martinez and T. Brandt Ryder, "The Problem with Paradigms: Bateman's Worldview as a Case Study," *Integrative and Comparative Biology* 45, no. 5 (2005): 821–830, https://doi.org/10.1093/icb/45.5.821; Ambika Kamath and Ashton Wesner, "Animal Territoriality, Property and Access: A Collaborative Exchange between Animal Behaviour and the Social Sciences," *Animal Behaviour* (2020): https://doi.org/10.1016/j.anbehav.2019.12.009.

9. Ambika Kamath and Jonathan Losos, "The Erratic and Contingent Progression of Research on Territoriality: A Case Study," *Behavioral Ecology and Sociobiology* 71, no. 6 (2017): 89, https://doi.org/10.1007/s00265-017-2319-z.

10. G. K. Noble and H. T. Bradley, "The Mating Behavior of Lizards; Its Bearing on the Theory of Sexual Selection," *Annals of the New York Academy of Sciences* 35, no. 1 (1933): 25–100, https://doi.org/10.1111/j.1749-6632.1933.tb55365.x.

11. Fred G. Thompson, "Notes on the Behavior of the Lizard *Anolis carolinensis*," *Copeia* 1954, no. 4 (1954): 299, https://doi.org/10.2307/1440053.

12. Aph Ko, *Racism as Zoological Witchcraft: A Guide for Getting Out* (Brooklyn: Lantern Books, 2019).

13. See, for example, Combahee River Collective, "A Black Feminist Statement" (Boston, 1977), https://combaheerivercollective.weebly.com/the-combahee-river-collective-statement.html; Judith Butler, *Gender Trouble: Feminism and the Subversion of Identity* (New York: Routledge, 1990); Kimberlé Crenshaw et al., *Critical Race Theory: The Key Writings That Formed the Movement* (New York: The New Press, 1995); María Lugones, "Heterosexualism and the Colonial/Modern Gender System," *Hypatia* 22, no. 1 (2007): 186–219, https://doi.org/10.1111/j.1527-2001.2007.tb01156.x.

14. Gregg Mitman, *The State of Nature: Ecology, Community, and American Social Thought, 1900–1950* (Chicago: University of Chicago Press, 1992), 9.

15. Gail Bederman, *Manliness & Civilization: A Cultural History of Gender and Race in the United States, 1880–1917* (Chicago: University of Chicago Press, 1995); Kamath and Wesner, "Animal Territoriality, Property and Access."

16. For example: Irina Pettersson and Cecilia Berg, "Environmentally Relevant Concentrations of Ethynylestradiol Cause Female-Biased Sex Ratios in *Xenopus tropicalis* and *Rana temporaria*," *Environmental Toxicology and Chemistry* 26, no. 5 (2007): 1005–1009, https://doi.org/10.1897/06-464r.1; Tyrone B. Hayes et al., "Hermaphroditic, Demasculinized Frogs after Exposure to the Herbicide Atrazine at Low Ecologically Relevant Doses," *Proceedings of the National Academy of Sciences* 99, no. 8 (2002): 5476–5480, https://doi.org/10.1073/pnas.082121499.

17. Interview with Max Lambert, June 29, 2023.

18. For instance: Tyrone B. Hayes et al., "Atrazine Induces Complete Feminization and Chemical Castration in Male African Clawed Frogs (*Xenopus laevis*)," *Proceedings of the National Academy of Sciences* 107, no. 10 (2010): 4612–4617, https://doi.org/10.1073/pnas.0909519107. For additional references, see works cited in Melina Packer and Max R. Lambert, "What's

Gender Got to Do with It? Dismantling the Human Hierarchies in Evolutionary Biology and Environmental Toxicology for Scientific and Social Progress," *American Naturalist* 200, no. 1 (2022): 114–128, https://doi.org/10.1086/720131.

19. Max R. Lambert et al., "Suburbanization, Estrogen Contamination, and Sex Ratio in Wild Amphibian Populations," *Proceedings of the National Academy of Sciences* 112, no. 38 (2015): 11881–11886, https://doi.org/10.1073/pnas.1501065112; Max R. Lambert et al., "Molecular Evidence for Sex Reversal in Wild Populations of Green Frogs (*Rana clamitans*)," *PeerJ* 7 (2019): e6449, https://doi.org/10.7717/peerj.6449.

20. Max R. Lambert, Tariq Ezaz, and David K. Skelly, "Sex-Biased Mortality and Sex Reversal Shape Wild Frog Sex Ratios," *Frontiers in Ecology and Evolution* 9 (2021): 737, https://doi.org/10.3389/fevo.2021.756476.

21. Bob Connors, "Sex and the Suburban Transgender Frog," *NBC Connecticut* (blog), March 2, 2010, https://www.nbcconnecticut.com/news/local/those-are-some-confused-frogs/2061213/. For evidence of viable offspring from so-named intersex frogs, see Jussi S. Alho, Chikako Matsuba, and Juha Merilä, "Sex Reversal and Primary Sex Ratios in the Common Frog (*Rana temporaria*)," *Molecular Ecology* 19, no. 9 (2010): 1763–1773, https://doi.org/10.1111/j.1365-294X.2010.04607.x.

22. See, for example, the documentary *The Disappearing Male* (Eggplant Picture & Sound, Optix Digital Pictures, Red Apple Entertainment, 2008).

23. Packer and Lambert, "What's Gender Got to Do with It?"

24. For queer feminist approaches to endocrine disrupting chemicals, see Giovanna Di Chiro, "Polluted Politics? Confronting Toxic Discourse, Sex Panic, and Eco-Normativity," in *Queer Ecologies: Sex, Nature, Politics, Desire*, ed. Catriona Mortimer-Sandilands and Bruce Erickson (Bloomington: Indiana University Press, 2010), 199–230; Anne Pollock, "Queering Endocrine Disruption," in *Object-Oriented Feminism*, ed. Katherine Behar (Minneapolis: University of Minnesota Press, 2016), 183–199; and Alexis Shotwell, *Against Purity: Living Ethically in Compromised Times* (Minneapolis: University of Minnesota Press, 2016).

25. For example: Sarah Laskow, "The Sad Sex Lives of Suburban Frogs," *Atlas Obscura*, September 9, 2015, http://www.atlasobscura.com/articles/the-sad-sex-lives-of-suburban-frogs.

26. Interview with Lambert, June 29, 2023.

CHAPTER 2

1. Angus J. Bateman, "Intra-Sexual Selection in *Drosophila*," *Heredity* 2 (1948): 349–368.

2. It's worth mentioning that Bateman knew that some of these mutations affect mating behavior in a species of fruit fly closely related to the species he used in his experiment. Brian F. Snyder and Patricia Adair Gowaty, "A Reappraisal of Bateman's Classic Study of Intrasexual Selection," *Evolution* 61, no. 11 (2007): 2457–2468.

3. Charles Darwin, *The Descent of Man, and Selection in Relation to Sex* (London: J. Murray, 1871).

4. Tang-Martinez and Ryder, "The Problem with Paradigms"; Zuleyma Tang-Martínez, "Rethinking Bateman's Principles: Challenging Persistent Myths of Sexually Reluctant Females and Promiscuous Males," *Journal of Sex Research* 53, no. 4–5 (May 3, 2016): 532–559, https://doi.org/10.1080/00224499.2016.1150938.

5. Tang-Martinez and Ryder, "The Problem with Paradigms"; Tang-Martínez, "Rethinking Bateman's Principles."

6. Angus J. Bateman, "Intra-Sexual Selection in *Drosophila*," 361; Snyder and Gowaty, "A Reappraisal of Bateman's Classic Study of Intrasexual Selection," 2463.

7. Angus J. Bateman, "Intra-Sexual Selection in *Drosophila*," 364, 365.

8. Robert L. Trivers, "Parental Investment and Sexual Selection," in *Sexual Selection and the Descent of Man*, ed. Bernard Campbell (Chicago: Aldine, 1972), 35–57; Darwin, as cited in Hrdy, "Empathy, Polyandry, and the Myth of the Coy Female," 132.

9. Tang-Martínez, "Rethinking Bateman's Principles," 538.

10. Angela Saini, *Inferior: How Science Got Women Wrong—and the New Research That's Rewriting the Story* (Boston: Beacon Press, 2017). For instance, a 2016 paper synthesizing research on animal sexual behavior pronounced: "our study confirms conventional sex roles for [non-monogamous] species in accordance with the pioneering ideas by Darwin, Bateman, and Trivers. Sexual selection research over the last 150 years has not been carried out under false premises but instead is valid and provides a powerful explanation for differences between males and females." Tim Janicke et al., "Darwinian Sex Roles Confirmed across the Animal Kingdom," *Science Advances* 2, no. 2 (2016): e1500983, https://doi.org/10.1126/sciadv.1500983.

11. Tang-Martinez and Ryder, "The Problem with Paradigms"; Snyder and Gowaty, "A Reappraisal of Bateman's Classic Study of Intrasexual Selection."

12. Hanna Kokko and Johanna Mappes, "Multiple Mating by Females Is a Natural Outcome of a Null Model of Mate Encounters," *Entomologia Experimentalis et Applicata* 146, no. 1 (2013): 26–37, https://doi.org/10.1111/j.1570-7458.2012.01296.x.

13. Lucy Cooke, *Bitch: On the Female of the Species* (New York: Basic Books, 2022), 73.

14. Patricia Adair Gowaty, "Sexual Natures: How Feminism Changed Evolutionary Biology," *Signs: Journal of Women in Culture and Society* 28, no. 3 (2003): 916, https://doi.org/10.1086/345324.

15. Hrdy, "Empathy, Polyandry, and the Myth of the Coy Female," 135.

16. Tang-Martínez, "Rethinking Bateman's Principles," 825.

17. Hrdy, "Empathy, Polyandry, and the Myth of the Coy Female," 146.

18. Snyder and Gowaty, "A Reappraisal of Bateman's Classic Study of Intrasexual Selection"; Patricia Adair Gowaty, Yong-Kyu Kim, and Wyatt W. Anderson, "No Evidence of Sexual Selection in a Repetition of Bateman's Classic Study of *Drosophila melanogaster*,"

Proceedings of the National Academy of Sciences 109, no. 29 (2012): 11740–11745, https://doi.org/10.1073/pnas.1207851109.

19. Cooke, *Bitch*, 74.

20. Dustin R. Rubenstein and John Alcock, *Animal Behavior*, 11th ed. (Sunderland, MA: Sinauer Associates, 2019), 307.

21. Rubenstein and Alcock, *Animal Behavior*, 309.

22. Banu Subramaniam, "Snow Brown and the Seven Detergents: A Metanarrative on Science and the Scientific Method," *Women's Studies Quarterly* 28, no. 1/2 (2000): 296–304.

23. Donna Haraway, "Situated Knowledges: The Science Question in Feminism and the Privilege of Partial Perspective," *Feminist Studies* 14, no. 3 (1988): 575–599, https://doi.org/10.2307/3178066. Haraway was specifically contesting philosopher Thomas Nagel's claim that this "view from nowhere" was the proper perspective to adopt in matters of science and philosophy. See Thomas Nagel, *The View from Nowhere* (Oxford: Oxford University Press, 1989).

24. Haraway, "Situated Knowledges."

25. Donna Haraway was one of the more prolific feminist science studies authors analyzing how human social locations affect animal behavior research in particular. See, for example, Donna Haraway, *Primate Visions: Gender, Race, and Nature in the World of Modern Science* (New York: Routledge, 1989).

26. Jack D. Forbes, *Columbus and Other Cannibals: The Wétiko Disease of Exploitation, Imperialism, and Terrorism*, rev. ed. (New York: Seven Stories Press, 1978); Gregory Cajete, *Native Science: Natural Laws of Interdependence* (Santa Fe: Clear Light Publishers, 2000).

27. While his masculinism cannot be ignored, Frantz Fanon was arguably one of the most powerful and brilliant writers on the violence of colonial science. See, for example, Frantz Fanon, "Colonial Violence and Mental Disorders," in *The Wretched of the Earth*, trans. Richard Philcox (New York: Grove Press, 1961).

28. Susan Bordo, "The Cartesian Masculinization of Thought," *Signs* 11, no. 3 (1986): 439–456.

29. For more on this history, see, for example, Peder Anker, *Imperial Ecology: Environmental Order in the British Empire, 1895–1945* (Cambridge, MA: Harvard University Press, 2001); and Lisa Lowe, *The Intimacies of Four Continents* (Durham, NC: Duke University Press, 2015).

30. As cited in Suman Seth, "Darwin and the Ethnologists: Liberal Racialism and the Geological Analogy," *Historical Studies in the Natural Sciences* 46, no. 4 (2016): 490–527, https://doi.org/10.1525/hsns.2016.46.4.490.

31. David Graeber and David Wengrow, *The Dawn of Everything: A New History of Humanity* (New York: Farrar, Straus and Giroux, 2021).

32. Renny Thomas, "Brahmins as Scientists and Science as Brahmins' Calling: Caste in an Indian Scientific Research Institute," *Public Understanding of Science* (2020): 306–318, https://doi.org/10.1177/0963662520903690; Kelsey Jo Starr and Neha Sahgal, "Measuring Caste in

India," *Decoded* (blog), June 29, 2021, https://www.pewresearch.org/decoded/2021/06/29/measuring-caste-in-india/.

33. Thomas, "Brahmins as Scientists and Science as Brahmins' Calling."
34. M. Gadgil and K. C. Malhotra, "Adaptive Significance of the Indian Caste System: An Ecological Perspective," *Annals of Human Biology* 10, no. 5 (1983): 465–477, https://doi.org/10.1080/03014468300006671.
35. Thomas, "Brahmins as Scientists and Science as Brahmins' Calling," 313.
36. Personal communication, 2015.
37. For a great summary of feminist standpoint theory, which we paraphrase here, see T. Bowell, "Feminist Standpoint Theory," Internet Encyclopedia of Philosophy, n.d., https://iep.utm.edu/fem-stan/.
38. For early observations on this shift, see Sandra Harding, *Whose Science? Whose Knowledge? Thinking from Women's Lives* (Ithaca, NY: Cornell University Press, 1991).
39. Malin Ah-King, *The Female Turn: How Evolutionary Science Shifted Perceptions about Females* (Palgrave Macmillan, 2022).
40. Ruth Hubbard, "Science, Facts, and Feminism," *Hypatia* 3, no. 1 (1988): 14.
41. Kokko and Mappes, "Multiple Mating by Females Is a Natural Outcome of a Null Model of Mate Encounters," 34.
42. Indeed, other work from at least one of these authors, Hanna Kokko, is decidedly unfeminist in its framings and assumptions. For example, in a 2005 paper, titled "Treat 'em Mean, Keep 'em (Sometimes) Keen: Evolution of Female Preferences for Dominant and Coercive Males," Kokko sets up a mathematical model that assumes "that male dominance (or coerciveness) improves his mating success." Hanna Kokko, "Treat 'em Mean, Keep 'em (Sometimes) Keen: Evolution of Female Preferences for Dominant and Coercive Males," *Evolutionary Ecology* 19, no. 2 (2005): 123–135, https://doi.org/10.1007/s10682-004-7919-1.
43. Nagel, *The View from Nowhere*, 439.
44. Nagel, *The View from Nowhere*, 439.
45. Ed Yong, *An Immense World: How Animal Senses Reveal the Hidden Realms around Us* (New York: Random House, 2022), 6; original emphasis.
46. Yong, *An Immense World*, 141.
47. Yong, *An Immense World*, 354–355.
48. Anne Harrington, *Reenchanted Science: Holism in German Culture from Wilhelm II to Hitler* (Princeton, NJ: Princeton University Press, 1996), 55.
49. Harrington, *Reenchanted Science*, 59–60.
50. Harrington, *Reenchanted Science*, 38.
51. Harrington, *Reenchanted Science*, 62.

CHAPTER 3

1. Lori Marino, "Thinking Chickens: A Review of Cognition, Emotion, and Behavior in the Domestic Chicken," *Animal Cognition* 20, no. 2 (2017): 127–147, https://doi.org/10.1007/s10071-016-1064-4.

2. W. M. Muir, "Group Selection for Adaptation to Multiple-Hen Cages: Selection Program and Direct Responses," *Poultry Science* 75, no. 4 (1996): 447–458, https://doi.org/10.3382/ps.0750447

3. Peter J. Bowler, *Evolution: The History of an Idea*, 25th ann. ed. (Berkeley: University of California Press, 2009), 160.

4. W. M. Muir, "Group Selection for Adaptation to Multiple-Hen Cages: Selection Program and Direct Responses," 447. Muir actually selected not for the total number of eggs produced but for the product of egg number and egg weight (i.e., total egg weight). However, he found no changes over time in the weight of individual eggs, implying that changes in total egg weight were driven by changes in the number of eggs laid. For simplicity's sake, we refer to selection on, and changes in, egg number.

5. "When the Strong Outbreed the Weak: An Interview with William Muir—This View of Life," July 11, 2016, https://thisviewoflife.com/when-the-strong-outbreed-the-weak-an-interview-with-william-muir/.

6. J. V. Craig and W. M. Muir, "Group Selection for Adaptation to Multiple-Hen Cages: Beak-Related Mortality, Feathering, and Body Weight Responses," *Poultry Science* 75, no. 3 (1996): 294–302, https://doi.org/10.3382/ps.0750294; J. V. Craig and W. M. Muir, "Group Selection for Adaptation to Multiple-Hen Cages: Behavioral Responses," *Poultry Science* 75, no. 10 (1996): 1145–1155, https://doi.org/10.3382/ps.0751145.

7. "When the Strong Outbreed the Weak."

8. Marc Bekoff and Jessica Pierce, *Wild Justice: The Moral Lives of Animals* (Chicago: University of Chicago Press, 2010), 102–103.

9. Bekoff and Pierce, *Wild Justice*, 102–103.

10. Michael J. Wade, *Adaptation in Metapopulations: How Interaction Changes Evolution* (Chicago: University of Chicago Press, 2016).

11. Patrick Abbot et al., "Inclusive Fitness Theory and Eusociality," *Nature* 471, no. 7339 (2011): E1–E4, https://doi.org/10.1038/nature09831.

12. For more on the history and theory of multilevel selection, see Mitman, *The State of Nature*; David Sloan Wilson, *Does Altruism Exist? Culture, Genes, and the Welfare of Others* (New Haven, CT: Yale University Press, 2015); Wade, *Adaptation in Metapopulations*; and Samir Okasha, *Agents and Goals in Evolution* (Oxford: Oxford University Press, 2018).

13. In a conversation with David Sloan Wilson, Bill Muir described his commitment to free market capitalism as a direct result of his understanding of multilevel selection in animals: "We can learn much from animals. While we no longer evolve genetically, we have the

intellect to evolve socially and should take the lessons learned from genetic evolution and apply them to society for social evolution to advance. Multi-level selection is powerful, but requires safeguards to prevent cheating and breakdowns of that path. Ironically, between group (company) selection based on capitalism is the only way to keep the system honest. In a capitalistic society, those groups (companies) where cooperation fails will soon be out of business. This process is social selection at the group (company) level." "When the Strong Outbreed the Weak."

14. For example: Max Horkeimer and Theodor Adorno, "The Concept of Enlightenment," in *Dialectic of Enlightenment*, trans. John Cumming (New York: Herder & Herder, 1944); Susan Bordo, "The Cartesian Masculinization of Thought," *Signs* 11, no. 3 (1986): 439–456; Helen Longino, "Does the Structure of Scientific Revolutions Permit a Feminist Revolution in Science?," in *Thomas Kuhn*, ed. Thomas Nickles (New York: Cambridge University Press, 2002), 272, https://doi.org/10.1017/CBO9780511613975.
15. Rubenstein and Alcock, *Animal Behavior*, 1.
16. Rubenstein and Alcock, *Animal Behavior*, 1.
17. Jonathan Birch and Samir Okasha, "Kin Selection and Its Critics," *BioScience* 65, no. 1 (2015): 22–32, https://doi.org/10.1093/biosci/biu196.
18. Wilson, *Does Altruism Exist?*, 32.
19. Wilson, *Does Altruism Exist?*, 32.
20. David Sloan Wilson, "Altruism and Organism: Disentangling the Themes of Multi-level Selection Theory," *American Naturalist* 150, no. S1 (July 1997): S127, https://doi.org/10.1086/286053.
21. Wilson, *Does Altruism Exist?*, 51; Wade, *Adaptation in Metapopulations*, 160.
22. Marshall Sahlins, *The Use and Abuse of Biology: An Anthropological Critique of Sociobiology* (Ann Arbor: University of Michigan Press, 1977), 22.
23. For a small sliver of the vast variety of different historical and cultural conceptions of kin, see Helen Rose Ebaugh and Mary Curry, "Fictive Kin as Social Capital in New Immigrant Communities," *Sociological Perspectives* 43, no. 2 (2000): 189–209, https://doi.org/10.2307/1389793; Enrique Salmón, "Kincentric Ecology: Indigenous Perceptions of the Human-Nature Relationship," *Ecological Applications* 10, no. 5 (2000): 1327–1332; Sonia Ryang, "A Note on Transnational Consanguinity, or, Kinship in the Age of Terrorism," *Anthropological Quarterly* 77, no. 4 (2004): 747–770; Marilyn Strathern, *Kinship, Law and the Unexpected: Relatives Are Always a Surprise* (New York: Cambridge University Press, 2005); Cristian Alvarado Leyton, "Ritual and Fictive Kinship," in *The International Encyclopedia of Anthropology*, ed. Hilary Callan (Hoboken, NJ: Wiley, 2018), 1–3; and Eduardo Kohn, *How Forests Think: Toward an Anthropology beyond the Human* (Berkeley: University of California Press, 2013).
24. Sahlins, *The Use and Abuse of Biology*, 87.

25. Interview with Emily DuVal, July 29, 2023.

26. Emily H. DuVal, "Adaptive Advantages of Cooperative Courtship for Subordinate Male Lance-Tailed Manakins," *American Naturalist* 169, no. 4 (April 2007): 423–432, https://doi.org/10.1086/512137.

27. See, for example, Dean Spade, "Solidarity Not Charity: Mutual Aid for Mobilization and Survival," *Social Text* 38, no. 1 (142) (March 1, 2020): 131–151, https://doi.org/10.1215/01642472-7971139; and Leah Lakshmi Piepzna-Samarasinha, *Care Work: Dreaming Disability Justice* (Vancouver: Arsenal Pulp Press, 2018).

28. For instance, Black feminist theorist Saidiya Hartman has said that "slavery is the ghost in the machine of kinship," meaning that "it is not possible to separate questions of kinship from property relations (and conceiving persons as property) and from the fictions of 'bloodline,' as well as the national and racial interests by which these lines are sustained." As cited in Judith Butler, "Is Kinship Always Already Heterosexual?," *Differences* 13, no. 1 (May 1, 2002): 15, https://doi.org/10.1215/10407391-13-1-14. Also see Carol B. Stack, *All Our Kin: Strategies for Survival in a Black Community* (1974; repr., New York: Basic Books, 2003).

29. For more on chosen, queer, and multispecies kin, see Kath Weston, *Families We Choose: Lesbians, Gays, Kinship* (New York: Columbia University Press, 1997); and Harlan Weaver, "Pit Bull Promises: Inhuman Intimacies and Queer Kinships in an Animal Shelter," *GLQ: A Journal of Lesbian and Gay Studies* 21, no. 2 (2015): 343–363.

30. Alexis Pauline Gumbs, *Undrowned: Black Feminist Lessons from Marine Mammals* (Chico: AK Press, 2020), 162.

31. Mitman, *The State of Nature.*

32. Peter Kropotkin, *Mutual Aid: An Illuminated Factor of Evolution* (Toronto: PM Press, 2021), 35.

33. Kropotkin, *Mutual Aid*, 228.

34. W. C. Allee and Edith S. Bowen, "Studies in Animal Aggregations: Mass Protection against Colloidal Silver among Goldfishes," *Journal of Experimental Zoology* 61, no. 2 (1932): 185–207, https://doi.org/10.1002/jez.1400610202.

35. Mitman, *The State of Nature.*

36. Mitman, *The State of Nature*, 5.

37. Mitman, *The State of Nature*, 132.

38. As quoted in Mitman, *The State of Nature*, 164.

39. V. C. Wynne-Edwards, *Animal Dispersion in Relation to Social Behaviour* (Edinburgh: Oliver and Boyd, 1962); V. C. Wynne-Edwards, "Self-Regulating Systems in Populations of Animals: A New Hypothesis Illuminates Aspects of Animal Behavior That Have Hitherto Seemed Unexplainable," *Science* 147, no. 3665 (1965): 1543–1548, https://doi.org/10.1126/science.147.3665.1543.

40. Wynne-Edwards, "Self-Regulating Systems in Populations of Animals," 1548; V. C. Wynne-Edwards, "Population Control in Animals," *Scientific American* 211, no. 2 (1964): 74.

41. Wynne-Edwards, "Self-Regulating Systems in Populations of Animals," 1548.

42. Adryan Corcione, "Eco-Fascism: What It Is, Why It's Wrong, and How to Fight It," *Teen Vogue*, April 30, 2020, https://www.teenvogue.com/story/what-is-ecofascism-explainer. While outside the scope of this book, there is a long and ugly history of racist and totalitarian groups (mis)using environmental concerns—from "carrying capacity" to climate change—to justify eugenic policies and practices; the Nazis considered themselves environmentalists, for instance, while the founders of the US conservation movement were committed eugenicists (including national parks luminaries such as John Muir and Theodore Roosevelt). For more on these histories, see Dorceta E. Taylor, *The Rise of the American Conservation Movement: Power, Privilege, and Environmental Protection* (Durham, NC: Duke University Press, 2016); and Climate & Mind, "What Is the Psychology of Ecofascism? A Bibliography," Climate & Mind, 2019, https://www.climateandmind.org/what-is-the-psychology-of-ecofascism.

43. Charles C. Mann, "The Book That Incited a Worldwide Fear of Overpopulation," *Smithsonian Magazine*, accessed February 26, 2024, https://www.smithsonianmag.com/innovation/book-incited-worldwide-fear-overpopulation-180967499/. On the racial politics of birth control, see also Laura Briggs, *Reproducing Empire: Race, Sex, Science, and U.S. Imperialism in Puerto Rico* (Berkeley: University of California Press, 2003); and Michelle Murphy, *The Economization of Life* (Durham, NC: Duke University Press, 2017).

44. Ayelet Shavit, "Shifting Values Partly Explain the Debate over Group Selection," *Studies in History and Philosophy of Science Part C: Studies in History and Philosophy of Biological and Biomedical Sciences* 35, no. 4 (2004): 705, https://doi.org/10.1016/j.shpsc.2004.09.007.

45. Writing a new preface to this book thirty years later, Williams recalled the initial motivation for his book: a lecture he listened to by University of Chicago ecologist A. E. Emerson on the notion of the "beneficial death" of individuals for the good of the species: "my reaction was that if Emerson's presentation was acceptable biology, I would prefer another calling." George C Williams, *Adaptation and Natural Selection: A Critique of Some Current Evolutionary Thought*, Princeton Science Library ed. (Princeton, NJ: Princeton University Press, 1996), ix.

46. George C. Williams, "A Sociobiological Expansion of Evolution and Ethics," in *Evolution and Ethics: T.H. Huxley's Evolution and Ethics with New Essays on Its Victorian and Sociobiological Context*, ed. George Christopher Williams and James G. Paradis (Princeton, NJ: Princeton University Press, 1989), 196; Shavit, "Shifting Values Partly Explain the Debate over Group Selection."

47. Shavit, "Shifting Values Partly Explain the Debate over Group Selection," 78.

48. Edward O. Wilson, *Sociobiology: The New Synthesis* (Cambridge, MA: Belknap Press of Harvard University Press, 1975), 3.

49. Richard Dawkins, *The Selfish Gene* (Oxford: Oxford University Press, 1989), 47.

50. The default narrative of animal aggression is everywhere in animal behavior science. When Ambika realized that she too defaulted to attributing an aggressive intent to most communication between animals, especially between males, she began asking herself, when watching animals interact, "If I imagined people doing this, would I think that they were fighting or would I think they were just having a conversation whose content I can't quite decipher?" The answer, once you allow for it, is almost always the latter.
51. Dawkins, *The Selfish Gene*, 4.
52. As quoted in Wilson, *Does Altruism Exist?*, 32.
53. Williams, *Adaptation and Natural Selection*, xi.
54. Bernd Heinrich, *Ravens in Winter* (New York: Summit Books, 1989), 12.

CHAPTER 4

1. Keith D. Waddington and Larry R. Holden, "Optimal Foraging: On Flower Selection by Bees," *American Naturalist* 114, no. 2 (1979): 179–196, https://doi.org/10.1086/283467; Allen R. Lewis, "Selection of Nuts by Gray Squirrels and Optimal Foraging Theory," *American Midland Naturalist* 107, no. 2 (1982): 250–257, https://doi.org/10.2307/2425376; James C. Munger, "Optimal Foraging? Patch Use by Horned Lizards (Iguanidae: *Phrynosoma*)," *American Naturalist* 123, no. 5 (1984): 654–680, https://doi.org/10.1086/284230.
2. Waddington and Holden, "Optimal Foraging"; Lewis, "Selection of Nuts by Gray Squirrels and Optimal Foraging Theory"; Munger, "Optimal Foraging? Patch Use by Horned Lizards (Iguanidae: *Phrynosoma*)."
3. Rubenstein and Alcock, *Animal Behavior*, 211.
4. G. J. Pierce and J. G. Ollason, "Eight Reasons Why Optimal Foraging Theory Is a Complete Waste of Time," *Oikos* 49, no. 1 (1987): 114, https://doi.org/10.2307/3565560.
5. Okasha, *Agents and Goals in Evolution*.
6. Robert Skidelsky, *What's Wrong with Economics? A Primer for the Perplexed* (New Haven, CT: Yale University Press, 2020).
7. Skidelsky, *What's Wrong with Economics?*, 10.
8. Robert H. MacArthur and Eric R. Pianka, "On Optimal Use of a Patchy Environment," *American Naturalist* 100, no. 916 (1966): 603–609, https://doi.org/10.1086/282454.
9. Elizabeth Popp Berman, *Thinking like an Economist: How Efficiency Replaced Equality in U.S. Public Policy* (Princeton, NJ: Princeton University Press, 2022).
10. Søren Mau, *Mute Compulsion: A Marxist Theory of the Economic Power of Capital* (London: Verso, 2023), 7.
11. "When the Strong Outbreed the Weak."
12. Paul Krugman and Robin Wells, *Microeconomics*, 2nd ed. (New York: Worth Publishers, 2009), 249–250.

13. In their 1985 book *The Dialectical Biologist*, Levins and Lewontin wrote: "Putting an optimization program into practice requires a general theory of optimality, which evolutionists have taken directly from the economics of capitalism. . . . In such theories," they continue, "the criterion of optimality is efficiency, whether of time or invested energy, yet the moralistic and ideological overtones of 'efficiency,' 'waste,' 'maximum return on investment,' and 'best use of time' seem never to have come to the consciousness of evolutionists, who adhere to these social norms unquestioningly." Richard Levins and Richard C. Lewontin, *The Dialectical Biologist* (Cambridge, MA: Harvard University Press, 1985), 26.
14. Neoliberalism, in brief, is an extremely "free market" form of governance, demanding little to no government regulation under the premise that economic competition will necessarily produce the most effective and efficient result. Much has been written about neoliberalism; for some recent, and feminist, critiques, see Lisa Duggan, *The Twilight of Equality? Neoliberalism, Cultural Politics, and the Attack on Democracy* (Boston: Beacon Press, 2014); and Wendy Brown, *In the Ruins of Neoliberalism: The Rise of Antidemocratic Politics in the West* (New York: Columbia University Press, 2019).
15. Berman, *Thinking like an Economist.*
16. Berman, *Thinking like an Economist*, chap. 1.
17. Berman, *Thinking like an Economist*, chap. 10.
18. MacArthur and Pianka, "On Optimal Use of a Patchy Environment."
19. Levins and Lewontin, *The Dialectical Biologist*, 76.
20. Sahlins, *The Use and Abuse of Biology*, 75.
21. Michael Baym (@baym), "'Survival of the Fit Enough Where Fit Means Not Dying' or Even 'Survival of Those Which Don't Die' Is Unsatisfyingly Tautological, but It's Really What That Phrase Comes down To," *Twitter*, December 18, 2018, 12:55 p.m., https://twitter.com/baym/status/1075087112880488450.
22. Bowler, *Evolution*, 104.
23. Thomas Hobbes, *Leviathan: Or the Matter, Forme and Power of a Commonwealth, Ecclesiasticall and Civil* (London: Andrew Crooke, 1651).
24. Richard C. Lewontin and Richard Levins, *Biology under the Influence: Dialectical Essays on Ecology, Agriculture, and Health* (New York: Monthly Review Press, 2007), 60; Sahlins, *The Use and Abuse of Biology*, 93.
25. William Irvine, *Apes, Angels, and Victorians; the Story of Darwin, Huxley, and Evolution* (New York: McGraw-Hill, 1955), 98.
26. Sahlins, *The Use and Abuse of Biology*, 101–102.
27. Thomas S. Kuhn, *The Structure of Scientific Revolutions*, 2nd ed. (Chicago: University of Chicago Press, 1970), vii.
28. Kuhn, *The Structure of Scientific Revolutions*, 36, 35.

29. Science studies scholars call this "problem closure." Maarten A. Hajer, *The Politics of Environmental Discourse: Ecological Modernization and the Policy Process* (Oxford: Oxford University Press, 1995).

30. To drive the point home a bit further, consider this: the *Animal Behavior* textbook uses the phrase "Darwinian puzzle" to describe a diversity of research questions in the field a not-insubstantial twenty-three times. Rubenstein and Alcock, *Animal Behavior*, I–8.

31. Kuhn, *The Structure of Scientific Revolutions*, 53.

32. Kuhn, *The Structure of Scientific Revolutions*, 78.

33. Kuhn, *The Structure of Scientific Revolutions*, 4.

34. For a similar discussion of the persistence of scientific paradigms in relation to binary sex categorization, see Beans Velocci, "Binary Logic" (New Haven, CT: Yale University, 2021), 25–26.

35. Longino, "Does the Structure of Scientific Revolutions Permit a Feminist Revolution in Science?," 273.

36. Emilio J. Castilla, "Gender, Race, and Meritocracy in Organizational Careers," *American Journal of Sociology* 113, no. 6 (2008): 1479–1526, https://doi.org/10.1086/588738.

37. Stuart Hall, "Gramsci's Relevance for the Study of Race and Ethnicity," *Journal of Communication Inquiry* (1986): 12; Cedric J. Robinson, *Black Marxism: The Making of the Black Radical Tradition* (Chapel Hill: University of North Carolina Press, 2000); Destin Jenkins and Justin Leroy, eds., *Histories of Racial Capitalism* (New York: Columbia University Press, 2021).

38. We borrow the phrase "interlocking systems of oppression" from the Black feminist authors of the Combahee River Collective Statement. Combahee River Collective, "A Black Feminist Statement" (Boston, 1977), https://combaheerivercollective.weebly.com/the-combahee-river-collective-statement.html.

39. "Definitions | Economics | Vancouver Island University | Canada," accessed October 20, 2023, https://management.viu.ca/economics/definitions.

40. Robin Wall Kimmerer, "The Serviceberry: An Economy of Abundance," *Emergence Magazine*, October 26, 2022, https://emergencemagazine.org/essay/the-serviceberry/.

41. "The Serviceberry."

42. You might notice that, by centering mutual relationships, Kimmerer's worldview is more compatible with multilevel selection than individual-level natural selection (see chapter 3). Indeed, Kimmerer has reminded us, "such communal generosity might seem incompatible with the process of evolution, which invokes the imperative of individual survival. But we make a grave error if we try to separate individual well-being from the health of the whole." Robin Wall Kimmerer, *Braiding Sweetgrass: Indigenous Wisdom, Scientific Knowledge, and the Teachings of Plants* (Minneapolis: Milkweed Editions, 2013), 16.

43. "The Serviceberry"; Robin Wall Kimmerer, "The Covenant of Reciprocity," in *The Wiley Blackwell Companion to Religion and Ecology*, ed. John Hart (Hoboken, NJ: John Wiley & Sons, 2017), 368–381, https://doi.org/10.1002/9781118465523.ch26.
44. "The Serviceberry."
45. Kimmerer, *Braiding Sweetgrass*, 29.
46. Joan Roughgarden, *Evolution's Rainbow: Diversity, Gender, and Sexuality in Nature and People*, 10th ann. ed. (Berkeley: University of California Press, 2013), 1.
47. Nathan W. Bailey and Marlene Zuk, "Same-Sex Sexual Behavior and Evolution," *Trends in Ecology & Evolution* 24, no. 8 (2009): 442, https://doi.org/10.1016/j.tree.2009.03.014.
48. Savanna R. T. Boutin et al., "Same-Sex Sexual Behaviour in Crickets: Understanding the Paradox," *Animal Behaviour* 114 (April 1, 2016): 101–110, https://doi.org/10.1016/j.anbehav.2016.01.022.
49. Since we have no way of knowing whether other animals experience anything like our human sense of gender, or, if they do, what their experiences are, we prefer the term "nonbinary" instead of "trans" in reference to animals. "Nonbinary" not only avoids the pitfalls of projecting human concepts of gender onto animals, but also better describes sex (and gender) as a spectrum, rather than as an either/or. Even "trans" can imply female *or* male, whereas "nonbinary" encapsulates the entire continuum.
50. Bruce Bagemihl, *Biological Exuberance: Animal Homosexuality and Natural Diversity*, Stonewall Inn ed. (New York: St. Martin's Press, 2000); Eliot Schrefer, *Queer Ducks (and Other Animals): The Natural World of Animal Sexuality* (New York: Katherine Tegen Books, 2022).
51. Optimality thinking also allows for the alternative possibility that other interlinked biological factors might somehow prevent natural selection from eliminating same-sex sexual behavior, despite the behavior's costs.
52. Bailey and Zuk, "Same-Sex Sexual Behavior and Evolution"; Bagemihl, *Biological Exuberance*, 106–115.
53. Bagemihl, *Biological Exuberance*, 107.
54. As quoted in Bagemihl, *Biological Exuberance*, 107. Also see this interview with Eliot Schrefer on *The Daily Show* with Trevor Noah: "Eliot Schrefer—Queer Ducks (and Other Animals)," YouTube, June 10, 2022, https://www.youtube.com/watch?v=jJRp8lEAUBU.
55. Catriona Mortimer-Sandilands and Bruce Erickson, eds., *Queer Ecologies: Sex, Nature, Politics, Desire* (Bloomington: Indiana University Press, 2010).
56. Nicole Seymour, *Strange Natures: Futurity, Empathy, and the Queer Ecological Imagination* (Urbana: University of Illinois Press, 2013), 32.
57. Julia D. Monk et al., "An Alternative Hypothesis for the Evolution of Same-Sex Sexual Behaviour in Animals," *Nature Ecology & Evolution* (2019): 6, https://doi.org/10.1038/s41559-019-1019-7.

58. Of course, in specific cases where individuals only engage in same-sex sexual behavior, it's quite possible that these individuals will not reproduce. In practice, the likelihood of this outcome depends on precisely how researchers are classifying individuals into sexes; as we saw in the example of the frogs in chapter 1, individuals that are classified as different sexes at different points in their lives are nonetheless capable of producing offspring.

59. Inon Scharf and Oliver Y. Martin, "Same-Sex Sexual Behavior in Insects and Arachnids: Prevalence, Causes, and Consequences," *Behavioral Ecology and Sociobiology* 67, no. 11 (2013): 1719–1730, https://doi.org/10.1007/s00265-013-1610-x; Chang S. Han and Robert C. Brooks, "Same-Sex Sexual Behaviour as a by-Product of Reproductive Strategy under Male–Male Scramble Competition," *Animal Behaviour* 108 (2015): 193–197, https://doi.org/10.1016/j.anbehav.2015.07.035.

60. Jackson Clive, Ewan Flintham, and Vincent Savolainen, "Same-Sex Sociosexual Behaviour Is Widespread and Heritable in Male Rhesus Macaques," *Nature Ecology & Evolution* 7, no. 8 (2023): 1287–1301, https://doi.org/10.1038/s41559-023-02111-y.

61. Monk et al., "An Alternative Hypothesis for the Evolution of Same-Sex Sexual Behaviour in Animals."

62. Bagemihl, *Biological Exuberance*, 209.

63. Michael Baym (@baym), "'Survival of the Fit Enough Where Fit Means Not Dying' or Even 'Survival of Those Which Don't Die' Is Unsatisfyingly Tautological, but It's Really What That Phrase Comes down To."

64. A funny and also telling observation: the idea of the "survival of the fittest" is so deeply embedded in dominant culture that, at the time of writing, Google Docs' artificial-intelligence-powered language-correction tool suggested that we edit the phrase "survival of the fit enough" to read—you guessed it—"survival of the fittest."

65. Tithi Bhattacharya, "Introduction: Mapping Social Reproduction Theory," in *Social Reproduction Theory: Remapping Class, Recentering Oppression*, ed. Tithi Bhattacharya (London: Pluto Press, 2017), 1–2.

66. Sandra Harding, "'Strong Objectivity': A Response to the New Objectivity Question," *Synthese* 104, no. 3 (September 1, 1995): 331–349, https://doi.org/10.1007/BF01064504.

67. Chanda Prescod-Weinstein and Katherine McKittrick, "Public Thinker: Katherine McKittrick on Black Methodologies and Other Ways of Being," *Public Books* (blog), February 1, 2021, https://www.publicbooks.org/public-thinker-katherine-mckittrick-on-black-methodologies-and-other-ways-of-being/.

68. Ambika first came across a similar metaphor in Levins and Lewontin, *The Dialectical Biologist*, 271.

69. A version of this image has been floating around on the internet for several years; a reverse image search leads to the oldest usage on Imgur in 2016 with the title "No, things aren't that simple," where, amazingly, this image was part of a more complicated context that added a third perspective; subsequent users might have cropped the image to what is shown here.

Imgur, "No, Things Aren't That Simple.—Post," Imgur, accessed October 19, 2023, https://imgur.com/gallery/1zZ6VSe.

70. See, for example, Whitney N. Laster Pirtle, "Racial Capitalism: A Fundamental Cause of Novel Coronavirus (COVID-19) Pandemic Inequities in the United States," *Health Education & Behavior* 47, no. 4 (August 1, 2020): 504–508, https://doi.org/10.1177/1090198120922942.

71. Even the ways that scholars talk and write about this co-constitution is multiple and varied! Karen Barad uses the term "intra-action" for this process; Donna Haraway prefers "becoming with," while feminist science historian Murphy offers "entanglements." We'll use "co-constitution" throughout our text. Karen Michelle Barad, *Meeting the Universe Halfway: Quantum Physics and the Entanglement of Matter and Meaning* (Durham, NC: Duke University Press, 2007); Donna J. Haraway, *When Species Meet* (Minneapolis: University of Minnesota Press, 2008); Michelle Murphy, *Sick Building Syndrome and the Problem of Uncertainty: Environmental Politics, Technoscience, and Women Workers* (Durham, NC: Duke University Press, 2006).

72. The Lakota/Dakota people are also known as Oceti Sakowin Oyate, or the People of the Seven Council Fires. Kimberly TallBear, "Why Interspecies Thinking Needs Indigenous Standpoints," *Cultural Anthropology*, April 24, 2011, https://culanth.org/fieldsights/260-why-interspecies-thinking-needs-indigenous-standpoints.

73. TallBear, "Why Interspecies Thinking Needs Indigenous Standpoints."

CHAPTER 5

1. Jonathan Losos, *Lizards in an Evolutionary Tree: Ecology and Adaptive Radiation of Anoles* (Oakland: University of California Press, 2011).

2. Interview with Jonathan Losos, May 16, 2023.

3. Losos has informally recruited pretty much every lizard biologist he knows (including Ambika) into this project of gathering observations of lizards with missing limbs, which explains how this dataset includes individuals from at least ten different lizard families, and from across the globe. He's promised us all coauthorship, when these data are eventually published as a scientific paper.

4. Interview with Jonathan Losos, May 16, 2023.

5. Interview with Jonathan Losos, May 16, 2023.

6. Haley A. Branch et al., "Discussions of the 'Not So Fit': How Ableism Limits Diverse Thought and Investigative Potential in Evolutionary Biology," *American Naturalist* 200, no. 1 (2022): 101–113, https://doi.org/10.1086/720003.

7. Alison Kafer, *Feminist, Queer, Crip* (Bloomington: Indiana University Press, 2013), 83.

8. See also Michele Friedner and Karen Weingarten, "Disability as Diversity: A New Biopolitics," *Somatosphere: Science, Medicine and Anthropology*, May 23, 2016, http://somatosphere.net/2016/05/disability-as-diversity-a-new-biopolitics.html.

9. Piepzna-Samarasinha, *Care Work*.

10. As quoted in Eli Clare, "Meditations on Natural Worlds, Disabled Bodies, and a Politics of Cure," in *Material Ecocriticism*, ed. Serenella Iovino and Serpil Oppermann (Bloomington: Indiana University Press, 2014), 210. See also Sunaura Taylor, *Beasts of Burden: Animal and Disability Liberation* (New York: New Press, 2017).

11. Black feminist geographer Ruth Wilson Gilmore famously defined racism as "the state-sanctioned or extralegal production and exploitation of group-differentiated vulnerability to premature death." Ruth Wilson Gilmore, *Golden Gulag: Prisons, Surplus, Crisis, and Opposition in Globalizing California* (Berkeley: University of California Press, 2007), 28.

12. As cited in Branch et al., "Discussions of the 'Not So Fit,'" 103; emphasis added.

13. Jessica Riskin, *The Restless Clock: A History of the Centuries-Long Argument over What Makes Living Things Tick* (Chicago: University of Chicago Press, 2018), 84; emphasis added.

14. Ernst Mayr, "How to Carry Out the Adaptationist Program?," *American Naturalist* 121, no. 3 (March 1983): 324, https://doi.org/10.1086/284064.

15. Levins and Lewontin, *The Dialectical Biologist*, 76.

16. Stephen Jay Gould and Richard C. Lewontin, "The Spandrels of San Marco and the Panglossian Paradigm: A Critique of the Adaptationist Programme," *Proceedings of the Royal Society of London. Series B: Biological Sciences* 205, no. 1161 (1979): 581–598, https://doi.org/10.1098/rspb.1979.0086.

17. Mayr, "How to Carry Out the Adaptationist Program?," 326.

18. Elisabeth A. Lloyd, "Adaptationism and the Logic of Research Questions: How to Think Clearly about Evolutionary Causes," *Biological Theory* 10, no. 4 (2015): 346; original emphasis, https://doi.org/10.1007/s13752-015-0214-2.

19. Historian Beans Velocci wrote about sex categorization in hyenas as a quintessential example of how late nineteenth- and early twentieth-century scientists used the incoherence of binary sex categorization to establish themselves as authorities on the natural world, a subject we get to in chapter 7. One scientist in particular, Morrison Watson, first insisted that it was impossibly difficult to distinguish male and female hyenas but then claimed that he was expert enough to distinguish them nonetheless. As Velocci put it, "in order for science to be the hero of the story, the hyena, especially the female hyena, had to first be constructed as confusing, and its sex as ambiguous" before scientists could swoop in to reinforce sex binaries through their expertise. Velocci, "Binary Logic," 59.

20. Rubenstein and Alcock, *Animal Behavior*, 278–279.

21. Rubenstein and Alcock, *Animal Behavior*, 279.

22. C. M. Drea et al., "Exposure to Naturally Circulating Androgens during Foetal Life Incurs Direct Reproductive Costs in Female Spotted Hyenas, but Is Prerequisite for Male Mating," *Proceedings of the Royal Society of London. Series B: Biological Sciences* 269, no. 1504 (2002): 1981–1987, https://doi.org/10.1098/rspb.2002.2109.

23. The original argument for the costliness of hyena clitorises in the wild was based on a comparison of mortality rates of hyenas and lions. While lions certainly do have smaller clitorises than hyenas, they are also entirely different animals; comparing lions to hyenas cannot explain why hyenas have large clitorises. Laurence G. Frank, Mary L. Weldele, and Stephen E. Glickman, "Masculinization Costs in Hyaenas," *Nature* 377, no. 6550 (1995): 584–585, https://doi.org/10.1038/377584b0.

24. Levins and Lewontin, *The Dialectical Biologist*, 79.

25. Evolutionary convergence, of course, is also not all or nothing. Notwithstanding their similarities, aquatic animals are *not* identical! An adaptationist worldview likely leads evolutionary biologists to overestimate the degree of convergence across animals that share not only some aspects of their environments but also some similar traits. For a specific example of how to study this, see Yoel E. Stuart et al., "Contrasting Effects of Environment and Genetics Generate a Continuum of Parallel Evolution," *Nature Ecology & Evolution* 1, no. 6 (2017): 1–7, https://doi.org/10.1038/s41559-017-0158.

26. Levins and Lewontin, *The Dialectical Biologist*, 99. The idea of organism-environment co-constitution is essentially identical to the concept of "niche construction," which was proposed (without that name) by Levins and Lewontin in *The Dialectical Biologist*, if not earlier, as part of an explicitly Marxist analysis of adaptationism. Niche construction has recently sparked much debate amongst evolutionary biologists—is it a *new* way of looking at evolution, or simply the same as reciprocal causation? We see organism-environment co-constitution (and niche construction) as fundamentally different from reciprocal causation because the latter remains deeply reductionist. We also believe that the depoliticization of niche construction by more recent proponents of the idea (as described by Erik I. Svensson, "On Reciprocal Causation in the Evolutionary Process," *Evolutionary Biology* 45, no. 1 [2018]: 1–14, https://doi.org/10.1007/s11692-017-9431-x) has made it *less* likely to get taken seriously by evolutionary biologists, because said depoliticization allows more readily for the absorption of niche construction into the field's existing paradigms of adaptationism and optimality thinking.

27. For more on the biology of tent caterpillars, see Terence D. Fitzgerald, *The Tent Caterpillars* (Ithaca, NY: Cornell University Press, 1995).

28. William H. Sewell Jr., *Logics of History: Social Theory and Social Transformation* (Chicago: University of Chicago Press, 2005), 6.

29. Sewell Jr., *Logics of History*, 7; original emphasis.

30. Sewell Jr., *Logics of History*, 8.

31. This argument leans implicitly on the concept of *overdetermination*. For more on overdetermination, see Louis Althusser, "Contradiction and Overdetermination: Notes for an Investigation," in *For Marx*, trans. Ben Brewster (London: Verso Books, 2005); William Lewis, "Louis Althusser," in *The Stanford Encyclopedia of Philosophy*, ed. Edward N. Zalta and Uri Nodelman, Fall 2022 (Metaphysics Research Lab, Stanford University, 2022), https://plato

.stanford.edu/archives/fall2022/entries/althusser/; and Hall, "Gramsci's Relevance for the Study of Race and Ethnicity."

32. Eileen Crist, *Images of Animals: Anthropomorphism and Animal Mind* (Philadelphia: Temple University Press, 1999).

33. Quoted in Crist, *Images of Animals*, 198.

34. Quoted in Crist, *Images of Animals*, 118.

35. Frans B. M. de Waal, *Are We Smart Enough to Know How Smart Animals Are?* (New York: W. W. Norton, 2016), 25.

36. In a 2021 paper, evolutionary biologists Sonia Sultan, Armin Moczek, and Denis Walsh defined organismal agency as "the capacity of a system to participate in its own persistence, maintenance, and function by regulating its own structures in response to the conditions it encounters." Sonia E. Sultan, Armin P. Moczek, and Denis Walsh, "Bridging the Explanatory Gaps: What Can We Learn from a Biological Agency Perspective?," *BioEssays* 44, no. 1 (2022): 4, https://doi.org/10.1002/bies.202100185.

37. Frans B. M. de Waal, "Moral Behavior in Animals," TEDxPeachtree, November 2011, https://www.ted.com/talks/frans_de_waal_moral_behavior_in_animals/transcript.

38. Darren Incorvaia, "A Two-Ton Lifeguard That Saved a Young Pup," *New York Times*, February 7, 2024, sec. Science, https://www.nytimes.com/2024/02/07/science/elephant-seals-pup-drowning.html.

39. Tiffany P, "Crow Uses Plastic Lid to Sled Down Roof Over and Over Again," YouTube, August 31, 2014, https://www.youtube.com/watch?v=L9mrTdYhOHg.

40. Riskin, *The Restless Clock*, 251.

CHAPTER 6

1. For an account, albeit one steeped in narratives of intense competition, of dung beetles' dung rolling behavior, see Douglas J. Emlen, *Animal Weapons: The Evolution of Battle* (New York: Henry Holt, 2014).

2. Leigh W. Simmons and T. James Ridsdill-Smith, *Ecology and Evolution of Dung Beetles* (Oxford: John Wiley & Sons, 2011).

3. Marie Dacke et al., "Dung Beetles Use the Milky Way for Orientation," *Current Biology* 23, no. 4 (2013): 298–300, https://doi.org/10.1016/j.cub.2012.12.034.

4. For example, see "Wayfinding," accessed October 20, 2023, https://archive.hokulea.com/navigate/navigate.html.

5. de Waal, *Are We Smart Enough to Know How Smart Animals Are?*, 12.

6. Christoph Grüter and Walter M. Farina, "The Honeybee Waggle Dance: Can We Follow the Steps?," *Trends in Ecology & Evolution* 24, no. 5 (2009): 242–247, https://doi.org/10.1016/j

.tree.2008.12.007; Thomas H. Kunz and Wendy R. Hood, "Parental Care and Postnatal Growth in the Chiroptera," in *Reproductive Biology of Bats*, ed. Elizabeth G. Crichton and Philip H. Krutzsch (London: Academic Press, 2000), 415–468.

7. de Waal, *Are We Smart Enough to Know How Smart Animals Are?*, 12.
8. Grace Hussain, "The 30 Most Intelligent Animals in the World Might Surprise You," September 20, 2023, https://sentientmedia.org/which-animals-are-most-intelligent/.
9. Rubenstein and Alcock, *Animal Behavior*, 142.
10. Agassiz founded Harvard University's Museum of Comparative Zoology, where Ambika got her PhD.
11. Lulu Miller, *Why Fish Don't Exist: A Story of Loss, Love, and the Hidden Order of Life* (New York: Simon & Schuster, 2021), 25.
12. Miller, *Why Fish Don't Exist*, 50.
13. For more on this history and transition, see Kay Anderson, *Race and the Crisis of Humanism* (London: Routledge, 2006).
14. David Starr Jordan, *The Blood of the Nation: A Study of the Decay of Races Through Survival of the Unfit* (Boston: American Unitarian Association, 1902), 64, 65.
15. Bederman, *Manliness & Civilization*; Alexandra Minna Stern, *Eugenic Nation: Faults and Frontiers of Better Breeding in Modern America*, 2nd ed. (Berkeley: University of California Press, 2015); Dorceta E. Taylor, *The Rise of the American Conservation Movement: Power, Privilege, and Environmental Protection* (Durham, NC: Duke University Press, 2016).
16. Daniel J. Kevles, *In the Name of Eugenics: Genetics and the Uses of Human Heredity* (Cambridge, MA: Harvard University Press, 1995); Kyla Schuller, *The Biopolitics of Feeling: Race, Sex, and Science in the Nineteenth Century* (Durham, NC: Duke University Press, 2017).
17. Kevles, *In the Name of Eugenics*, 19.
18. "At the Center of Cure," in *Brilliant Imperfection: Grappling with Cure*, by Eli Clare (Durham, NC: Duke University Press, 2017), 103–123.
19. Macolm Harris, *Palo Alto: A History of California, Capitalism, and the World* (New York: Little, Brown, 2023), 66. Even more valuable than the faster horses themselves was the *potential* of faster horses locked up in the sperm of the best studs. In Stanford's eyes, the success of his scheme depended upon the traits of a winning horse being genetically determined and therefore inherited. We return to the subject of genetic determinism in chapter 7.
20. Jordan, *The Blood of the Nation*, 44–45.
21. The Eugenics Archives, "Intelligence and IQ Testing," Social Sciences and Humanities Research Council of Canada, 2015, https://www.eugenicsarchive.ca/encyclopedia?id=535eecb77095aa000000023a.
22. Harris, *Palo Alto*, 111.

23. Harris, *Palo Alto*, 110.

24. Runaway models of sexual selection are known as such because they involve the evolution of positive feedback loops between traits in one sex and the preference for that trait in the other sex, leading both traits to become more and more exaggerated—traits and preference "run away" with each other. Rubenstein and Alcock, *Animal Behavior*, 341.

25. Erika L. Milam, *Looking for a Few Good Males: Female Choice in Evolutionary Biology* (Baltimore: Johns Hopkins University Press, 2010), 164.

26. Milam, *Looking for a Few Good Males*, 47.

27. Major Leonard Darwin, "Mate Selection," *The Eugenics Review* 15, no. 3 (1923): 459–471.

28. Darwin, "Mate Selection," 459.

29. Darwin, "Mate Selection," 466.

30. Adding another layer of meaning to the contemporary concept of "white fragility," which refers to white people's general difficulty admitting to the persistence of white supremacy. See: Robin DiAngelo, "White Fragility," *The International Journal of Critical Pedagogy* 3, no. 3 (2011): 54–70.

31. As Leonard Darwin put it, arguing for eugenic intervention: "If the superior types are dying out, it is they who are now the biologically unfit." Darwin, "Mate Selection," 465.

32. Milam, *Looking for a Few Good Males*, 46.

CHAPTER 7

1. See, for example: Emily H. DuVal and Bart Kempenaers, "Sexual Selection in a Lekking Bird: The Relative Opportunity for Selection by Female Choice and Male Competition," *Proceedings of the Royal Society B: Biological Sciences* 275, no. 1646 (2008): 1995–2003, https://doi.org/10.1098/rspb.2008.0151; Rebecca J. Sardell, Bart Kempenaers, and Emily H. DuVal, "Female Mating Preferences and Offspring Survival: Testing Hypotheses on the Genetic Basis of Mate Choice in a Wild Lekking Bird," *Molecular Ecology* 23, no. 4 (2014): 933–946, https://doi.org/10.1111/mec.12652; E. H. DuVal and J. A. Kapoor, "Causes and Consequences of Variation in Female Mate Search Investment in a Lekking Bird," *Behavioral Ecology* 26, no. 6 (2015): 1537–1547, https://doi.org/10.1093/beheco/arv110; Blake Carlton Jones and Emily H. DuVal, "Mechanisms of Social Influence: A Meta-Analysis of the Effects of Social Information on Female Mate Choice Decisions," *Frontiers in Ecology and Evolution* 7 (2019): https://www.frontiersin.org/articles/10.3389/fevo.2019.00390.

2. Interview with Emily DuVal, July 19, 2023.

3. James Gorman, "Why Are Dogs So Friendly? The Answer May Be in 2 Genes," *New York Times*, July 19, 2017, sec. Science, https://www.nytimes.com/2017/07/19/science/dogs-genes-sociability.html; Ed Yong, "The Genes That Built a Home," *National Geographic*, January 16, 2013, https://www.nationalgeographic.com/science/article/genetics-burrowing.

4. Emily H. DuVal et al., "Inferred Attractiveness: A Generalized Mechanism for Sexual Selection That Can Maintain Variation in Traits and Preferences over Time," *PLOS Biology* 21, no. 10 (2023): e3002269, https://doi.org/10.1371/journal.pbio.3002269.
5. Aaron Panofsky, *Misbehaving Science: Controversy and the Development of Behavior Genetics* (Chicago: University of Chicago Press, 2014), 40.
6. Anderson, *Race and the Crisis of Humanism.*
7. Bowler, *Evolution*, 285.
8. Bowler, *Evolution*, 293.
9. See, for example, Clifton Crais and Pamela Scully, *Sara Baartman and the Hottentot Venus: A Ghost Story and a Biography* (Princeton, NJ: Princeton University Press, 2010).
10. Claire Jean Kim, *Dangerous Crossings: Race, Species, and Nature in a Multicultural Age* (Cambridge: Cambridge University Press, 2015), 18; original emphasis.
11. Panofsky, *Misbehaving Science*, 1.
12. Megan Molteni, "Buffalo Shooting Ignites a Debate over the Role of Genetics Researchers in White Supremacist Ideology," *STAT* (blog), May 23, 2022, https://www.statnews.com/2022/05/23/buffalo-shooting-ignites-debate-genetics-researchers-in-white-supremacist-ideology/.
13. Richard C. Lewontin, "An Estimate of Average Heterozygosity in Man," *American Journal of Human Genetics* 19, no. 5 (1967): 681–685; Stephen Jay Gould, *The Mismeasure of Man* (New York: W. W. Norton, 1996); Alondra Nelson, *The Social Life of DNA* (Boston: Beacon Press, 2016); Jenny Reardon, *Race to the Finish: Identity and Governance in an Age of Genomics* (Princeton, NJ: Princeton University Press, 2005); Dorothy E. Roberts, "Is Race-Based Medicine Good for Us? African American Approaches to Race, Biomedicine, and Equality," *Journal of Law, Medicine & Ethics* 36, no. 3 (2008): 537–545, https://doi.org/10.1111/j.1748-720X.2008.302.x; Kimberly TallBear, *Native American DNA: Tribal Belonging and the False Promise of Genetic Science* (Minneapolis: University of Minnesota Press, 2013).
14. Levins and Lewontin, *The Dialectical Biologist*, 119.
15. Ruth Hubbard, "The Political Nature of 'Human Nature,'" in *Feminist Frameworks: Alternative Theoretical Accounts of the Relationships Between Men and Women*, ed. Alison Jagger and Paula Rothenberg, 3rd ed. (New York: McGraw-Hill, 1993), 145.
16. Brenna M. Henn et al., "Why DNA Is No Key to Social Equality: On Kathryn Paige Harden's 'The Genetic Lottery,'" *Los Angeles Review of Books*, September 21, 2021, sec. Review, https://lareviewofbooks.org/article/why-dna-is-no-key-to-social-equality-on-kathryn-paige-hardens-the-genetic-lottery/.
17. C. Brandon Ogbunugafor, "DNA, Basketball, and Birthday Luck. A Review of *The Genetic Lottery: Why DNA Matters for Social Equality*," *American Journal of Biological Anthropology* 179, no. 3 (2022): 2, https://doi.org/10.1002/ajpa.24599. We use a shortened version of

Ogbunugafor's last name (Ogbunu) in the main text, as he tends to do in his public-facing writing.

18. Ogbunugafor, "DNA, Basketball, and Birthday Luck," 2.

19. Ogbunugafor, "DNA, Basketball, and Birthday Luck," 2.

20. David S. Moore and David Shenk, "The Heritability Fallacy," *WIREs Cognitive Science* 8, nos. 1–2 (2017): e1400, https://doi.org/10.1002/wcs.1400.

21. There are many fascinating books on the history of animal breeding and its intersections with racial, sexual, and class politics. See, for example, Harriet Ritvo, *The Animal Estate: The English and Other Creatures in the Victorian Age* (Cambridge, MA: Harvard University Press, 1987).

22. William Arkwright, *The Pointer and His Predecessors; an Illustrated History of the Pointing Dog from the Earliest Times* (London: A.L. Humphreys, 1902); Craig Koshyk, *Pointing Dogs, Volume Two: The British and Irish Breeds* (Winnipeg: Dog Willing, 2023).

23. Rob Boddice, *A History of Attitudes and Behaviours toward Animals in Eighteenth- and Nineteenth-Century Britain: Anthropocentrism and the Emergence of Animals* (Lewiston, NY: Edwin Mellen Press, 2008).

24. Taylor, *Beasts of Burden*, 39.

25. In many animal species, it is undeniable that environmental factors such as temperature play a role in the development of sexual traits; a determinist view (though not strictly a *biological determinist* view, as we're using the term here) would couch this as an example of sex being environmentally determined, whereas a view rooted in co-constitution and contingency would reject the possibility of determination at all.

26. A simple counterargument to this logic of females investing more than males in each fertilization: while a single egg is certainly larger than a single sperm cell, males release many sperm per ejaculation, and so the energy expended on fertilization by males and females is actually comparable. Note, however, that this counterargument remains within the logic of calculating costs and benefits, and that the feminist critique of these binarized stereotypes of males and females would be valid even if it turned out that producing eggs takes more energy than producing ejaculate. Emily Martin, "The Egg and the Sperm: How Science Has Constructed a Romance Based on Stereotypical Male-Female Roles," *Signs* 16, no. 3 (1991): 485–501.

27. Martin, "The Egg and the Sperm," 498–99.

28. Martin, "The Egg and the Sperm," 500.

29. Nelly Oudshoorn, *Beyond the Natural Body: An Archaeology of Sex Hormones* (New York: Routledge, 1994).

30. Oudshoorn, *Beyond the Natural Body*, 39.

31. Sarah S. Richardson, "Sexing the X: How the X Became the 'Female Chromosome,'" *Signs* 37, no. 4 (2012): 909–933, https://doi.org/10.1086/664477.

32. Deboleena Roy, *Molecular Feminisms: Biology, Becomings, and Life in the Lab* (Seattle: University of Washington Press, 2018), 99.

33. Katrina Karkazis et al., "Out of Bounds? A Critique of the New Policies on Hyperandrogenism in Elite Female Athletes," *American Journal of Bioethics* 12, no. 7 (July 1, 2012): 5, https://doi.org/10.1080/15265161.2012.680533.

34. Karkazis et al., "Out of Bounds?," 5–6.

35. Karkazis et al., "Out of Bounds?," 6.

36. Karkazis et al., "Out of Bounds?," 6.

37. Robert M. Sapolsky, *The Trouble with Testosterone: And Other Essays on the Biology of the Human Predicament* (New York: Touchstone, 1998); Lambert et al., "Molecular Evidence for Sex Reversal in Wild Populations of Green Frogs (*Rana clamitans*)"; Lambert, Ezaz, and Skelly, "Sex-Biased Mortality and Sex Reversal Shape Wild Frog Sex Ratios."

38. Sara E. Lipshutz and Kimberly A. Rosvall, "Neuroendocrinology of Sex-Role Reversal," *Integrative and Comparative Biology* 60, no. 3 (2020): 692–702, https://doi.org/10.1093/icb/icaa046; J. F. McLaughlin et al., "Multivariate Models of Animal Sex: Breaking Binaries Leads to a Better Understanding of Ecology and Evolution," *Integrative and Comparative Biology* 63, no. 4 (2023): 891–906, https://doi.org/10.1093/icb/icad027; Sapolsky, *The Trouble with Testosterone*.

39. For further examples, see Bagemihl, *Biological Exuberance*; Roughgarden, *Evolution's Rainbow*; Schrefer, *Queer Ducks (and Other Animals)*; and McLaughlin et al., "Multivariate Models of Animal Sex."

40. Velocci, "Binary Logic," 2.

41. See, for example, Thomas Laqueur, *Making Sex: Body and Gender from the Greeks to Freud* (Cambridge, MA: Harvard University Press, 1992); Oyèrónkẹ́ Oyěwùmí, *The Invention of Women: Making an African Sense of Western Gender Discourses* (Minneapolis: University of Minnesota Press, 1997); and Beatrice Medicine, "Directions in Gender Research in American Indian Societies: Two Spirits and Other Categories," *Online Readings in Psychology and Culture* 3, no. 1 (August 1, 2002), https://doi.org/10.9707/2307-0919.1024.

42. Judith Butler, "Subjects of Sex/Gender/Desire," in *Gender Trouble: Feminism and the Subversion of Identity*, Routledge Classics (New York: Routledge, 1990), 1–25.

43. As cited in D. A. Dewsbury, "The Darwin-Bateman Paradigm in Historical Context," *Integrative and Comparative Biology* 45, no. 5 (2005): 834, https://doi.org/10.1093/icb/45.5.831.

44. Kim, *Dangerous Crossings*, 18.

45. Sally Markowitz, "Pelvic Politics: Sexual Dimorphism and Racial Difference," *Signs* 26, no. 2 (2001): 398.

46. Velocci, "Binary Logic."

47. Velocci, "Binary Logic," 157–158; emphasis added.

48. Erum Salam, "Canada Issues Travel Advisory for LGBTQ+ Residents Visiting US," *The Guardian*, August 31, 2023, sec. World News, https://www.theguardian.com/world/2023/aug/31/canada-travel-warning-lgbtq-residents-us.

49. Justine Wadsack, Rachel Jones, and Cory McGarr, "Gender Transition; Prohibitions; Hormone Therapies," Pub. L. No. SB1702 (2023), https://www.azleg.gov/legtext/56leg/1R/bills/SB1702P.pdf.

50. This essay, from evolutionary anthropologist Agustín Fuentes, is a good example of such scholarly activism. Agustín Fuentes, "Here's Why Human Sex Is Not Binary," *Scientific American*, May 1, 2023, https://www.scientificamerican.com/article/heres-why-human-sex-is-not-binary/. See also Brian M. Donovan et al., "Sex and Gender Essentialism in Textbooks," *Science* 383, no. 6685 (February 23, 2024): 822–825, https://doi.org/10.1126/science.adi1188.

51. Miriam Miyagi, Eartha Mae Guthman, and Simón(e) Dow-Kuang Sun, "Transgender Rights Rely on Inclusive Language," *Science* 374, no. 6575 (2021): 1568–1569, https://doi.org/10.1126/science.abn3759. For an in-depth exploration of a "multimodal" model for sex, see McLaughlin et al., "Multivariate Models of Animal Sex."

52. Hane Htut Maung, "Classifying Sexes," *Journal of Diversity and Gender Studies* 10, no. 1 (2023): 35–52.

53. Maung, "Classifying Sexes," 46.

54. Feminist science studies scholar Sarah Richardson has also offered a similar concept of "sex contextualism." Sarah S. Richardson, "Sex Contextualism," *Philosophy, Theory, and Practice in Biology* 14, no. 2 (2022), https://doi.org/10.3998/ptpbio.2096.

55. Research institutions prohibit the release of study animals back into the wild, to avoid the spread of disease and other contamination. Wild-caught study animals are therefore often euthanized after experiments conclude.

56. Marc Bekoff and Jessica Pierce, *Wild Justice: The Moral Lives of Animals* (Chicago: University of Chicago Press, 2010).

57. We borrow this turn of phrase from feminist historian of science Rebecca Herzig's book of the same title (about the suffering of human scientists and experimental subjects). Rebecca M. Herzig, *Suffering for Science: Reason and Sacrifice in Modern America* (New Brunswick, NJ: Rutgers University Press, 2006).

58. Haraway, *When Species Meet*, 82; original emphasis.

59. Lori Gruen, *Entangled Empathy: An Alternative Ethic for Our Relationships with Animals* (New York: Lantern Books, 2015), 36.

60. Levins and Lewontin, *The Dialectical Biologist*, 257; emphasis added.

Bibliography

Abbot, Patrick, Jun Abe, John Alcock, Samuel Alizon, Joao A. C. Alpedrinha, Malte Andersson, Jean-Baptiste Andre, et al. "Inclusive Fitness Theory and Eusociality." *Nature* 471, no. 7339 (2011): E1–E4. https://doi.org/10.1038/nature09831.

Ah-King, Malin. *The Female Turn: How Evolutionary Science Shifted Perceptions about Females.* Singapore: Palgrave Macmillan, 2022.

Alho, Jussi S., Chikako Matsuba, and Juha Merilä. "Sex Reversal and Primary Sex Ratios in the Common Frog (*Rana temporaria*)." *Molecular Ecology* 19, no. 9 (2010): 1763–1773. https://doi.org/10.1111/j.1365-294X.2010.04607.x.

Allee, W. C., and Edith S. Bowen. "Studies in Animal Aggregations: Mass Protection against Colloidal Silver among Goldfishes." *Journal of Experimental Zoology* 61, no. 2 (1932): 185–207. https://doi.org/10.1002/jez.1400610202.

Althusser, Louis. "Contradiction and Overdetermination: Notes for an Investigation." In *For Marx*. Translated by Ben Brewster. London: Verso Books, 2005.

Anderson, Kay. *Race and the Crisis of Humanism*. London: Routledge, 2006.

Anker, Peder. *Imperial Ecology: Environmental Order in the British Empire, 1895–1945*. Cambridge, MA: Harvard University Press, 2001.

Arkwright, William. *The Pointer and His Predecessors; an Illustrated History of the Pointing Dog from the Earliest Times*. London: A. L. Humphreys, 1902.

Bagemihl, Bruce. *Biological Exuberance: Animal Homosexuality and Natural Diversity*. Stonewall Inn ed. New York: St. Martin's Press, 2000.

Bailey, Nathan W., and Marlene Zuk. "Same-Sex Sexual Behavior and Evolution." *Trends in Ecology & Evolution* 24, no. 8 (2009): 439–446. https://doi.org/10.1016/j.tree.2009.03.014.

Barad, Karen Michelle. *Meeting the Universe Halfway: Quantum Physics and the Entanglement of Matter and Meaning*. Durham, NC: Duke University Press, 2007.

Bateman, Angus J. "Intra-Sexual Selection in *Drosophila*." *Heredity* 2 (1948): 349–368.

Bederman, Gail. *Manliness & Civilization: A Cultural History of Gender and Race in the United States, 1880–1917*. Chicago: University of Chicago Press, 1995.

Bekoff, Marc, and Jessica Pierce. *Wild Justice: The Moral Lives of Animals*. Chicago: University of Chicago Press, 2010.

Berman, Elizabeth Popp. *Thinking like an Economist: How Efficiency Replaced Equality in U.S. Public Policy*. Princeton, NJ: Princeton University Press, 2022.

Bhattacharya, Tithi. "Introduction: Mapping Social Reproduction Theory." In *Social Reproduction Theory: Remapping Class, Recentering Oppression*, edited by Tithi Bhattacharya, 1–20. London: Pluto Press, 2017.

Birch, Jonathan, and Samir Okasha. "Kin Selection and Its Critics." *BioScience* 65, no. 1 (2015): 22–32. https://doi.org/10.1093/biosci/biu196.

Boddice, Rob. *A History of Attitudes and Behaviours toward Animals in Eighteenth- and Nineteenth-Century Britain: Anthropocentrism and the Emergence of Animals*. Lewiston, NY: Edwin Mellen Press, 2008.

Bordo, Susan. "The Cartesian Masculinization of Thought." *Signs* 11, no. 3 (1986): 439–456.

Boutin, Savanna R. T., Sarah J. Harrison, Lauren P. Fitzsimmons, Emily M. McAuley, and Susan M. Bertram. "Same-Sex Sexual Behaviour in Crickets: Understanding the Paradox." *Animal Behaviour* 114 (April 1, 2016): 101–110. https://doi.org/10.1016/j.anbehav.2016.01.022.

Bowell, T. "Feminist Standpoint Theory." Internet Encyclopedia of Philosophy, n.d. https://iep.utm.edu/fem-stan/.

Bowler, Peter J. *Evolution: The History of an Idea*. 25th ann. ed. Berkeley: University of California Press, 2009.

Branch, Haley A., Amanda N. Klingler, Kelsey J. R. P. Byers, Aaron Panofsky, and Danielle Peers. "Discussions of the 'Not So Fit': How Ableism Limits Diverse Thought and Investigative Potential in Evolutionary Biology." *American Naturalist* 200, no. 1 (2022): 101–113. https://doi.org/10.1086/720003.

Briggs, Laura. *Reproducing Empire: Race, Sex, Science, and U.S. Imperialism in Puerto Rico*. Berkeley: University of California Press, 2003.

Brown, Wendy. *In the Ruins of Neoliberalism: The Rise of Antidemocratic Politics in the West*. New York: Columbia University Press, 2019.

Butler, Judith. *Gender Trouble: Feminism and the Subversion of Identity*. New York: Routledge, 1990.

Butler, Judith. "Is Kinship Always Already Heterosexual?" *Differences* 13, no. 1 (May 1, 2002): 14–44. https://doi.org/10.1215/10407391-13-1-14.

Butler, Judith. "Subjects of Sex/Gender/Desire." In *Gender Trouble: Feminism and the Subversion of Identity*, 1–25. New York: Routledge, 1990.

Cajete, Gregory. *Native Science: Natural Laws of Interdependence*. Santa Fe: Clear Light Publishers, 2000.

Castilla, Emilio J. "Gender, Race, and Meritocracy in Organizational Careers." *American Journal of Sociology* 113, no. 6 (2008): 1479–1526. https://doi.org/10.1086/588738.

Clare, Eli. "At the Center of Cure." In *Brilliant Imperfection: Grappling with Cure*, 103–123. Durham, NC: Duke University Press, 2017.

Clare, Eli. "Meditations on Natural Worlds, Disabled Bodies, and a Politics of Cure." In *Material Ecocriticism*, edited by Serenella Iovino and Serpil Oppermann, 204–218. Bloomington: Indiana University Press, 2014.

Climate & Mind. "What Is the Psychology of Ecofascism? A Bibliography." Climate & Mind, 2019. https://www.climateandmind.org/what-is-the-psychology-of-ecofascism.

Clive, Jackson, Ewan Flintham, and Vincent Savolainen. "Same-Sex Sociosexual Behaviour Is Widespread and Heritable in Male Rhesus Macaques." *Nature Ecology & Evolution* 7, no. 8 (2023): 1287–1301. https://doi.org/10.1038/s41559-023-02111-y.

Combahee River Collective. "A Black Feminist Statement." Boston, 1977. https://combaheeriver collective.weebly.com/the-combahee-river-collective-statement.html.

Connors, Bob. "Sex and the Suburban Transgender Frog." *NBC Connecticut* (blog), March 2, 2010. https://www.nbcconnecticut.com/news/local/those-are-some-confused-frogs/2061213/.

Cooke, Lucy. *Bitch: On the Female of the Species*. New York: Basic Books, 2022.

Corcione, Adryan. "Eco-Fascism: What It Is, Why It's Wrong, and How to Fight It." *Teen Vogue*, April 30, 2020. https://www.teenvogue.com/story/what-is-ecofascism-explainer.

Craig, J. V., and W. M. Muir. "Group Selection for Adaptation to Multiple-Hen Cages: Beak-Related Mortality, Feathering, and Body Weight Responses." *Poultry Science* 75, no. 3 (1996): 294–302. https://doi.org/10.3382/ps.0750294.

Craig, J. V., and W. M. Muir. "Group Selection for Adaptation to Multiple-Hen Cages: Behavioral Responses." *Poultry Science* 75, no. 10 (1996): 1145–1155. https://doi.org/10.3382/ps.0751145.

Crais, Clifton, and Pamela Scully. *Sara Baartman and the Hottentot Venus: A Ghost Story and a Biography*. Princeton, NJ: Princeton University Press, 2010.

Crenshaw, Kimberlé, Neil Gotanda, Garry Peller, and Kendall Thomas. *Critical Race Theory: The Key Writings That Formed the Movement*. New York: The New Press, 1995.

Crist, Eileen. *Images of Animals: Anthropomorphism and Animal Mind*. Philadelphia: Temple University Press, 1999.

Dacke, Marie, Emily Baird, Marcus Byrne, Clarke H. Scholtz, and Eric J. Warrant. "Dung Beetles Use the Milky Way for Orientation." *Current Biology* 23, no. 4 (2013): 298–300. https://doi.org /10.1016/j.cub.2012.12.034.

Daily Show. "Eliot Schrefer—Queer Ducks (and Other Animals)." Interview by Trevor Noah. YouTube, June 10, 2022. https://www.youtube.com/watch?v=jJRp8lEAUBU.

Darwin, Charles. *On the Origin of Species by Means of Natural Selection*. London: J. Murray, 1859.

Darwin, Charles. *The Descent of Man, and Selection in Relation to Sex*. London: J. Murray, 1871.

Darwin, Major Leonard. “Mate Selection.” *The Eugenics Review* 15, no. 3 (1923): 459–471.

Dawkins, Richard. *The Selfish Gene*. Oxford: Oxford University Press, 1989.

“Definitions | Economics | Vancouver Island University | Canada.” Accessed October 20, 2023. https://management.viu.ca/economics/definitions.

de Waal, Frans B. M. *Are We Smart Enough to Know How Smart Animals Are?* New York: W. W. Norton, 2016.

de Waal, Frans B. M. “Moral Behavior in Animals.” TEDxPeachtree, November 2011. https://www.ted.com/talks/frans_de_waal_moral_behavior_in_animals/transcript.

Dewsbury, D. A. “The Darwin-Bateman Paradigm in Historical Context.” *Integrative and Comparative Biology* 45, no. 5 (2005): 831–837. https://doi.org/10.1093/icb/45.5.831.

Di Chiro, Giovanna. “Polluted Politics? Confronting Toxic Discourse, Sex Panic, and Eco-Normativity.” In *Queer Ecologies: Sex, Nature, Politics, Desire*, edited by Catriona Mortimer-Sandilands and Bruce Erickson, 199–230. Bloomington: Indiana University Press, 2010.

DiAngelo, Robin. “White Fragility.” *The International Journal of Critical Pedagogy* 3, no. 3 (2011): 54–70.

Donovan, Brian M., Awais Syed, Sophie H. Arnold, Dennis Lee, Monica Weindling, Molly A. M. Stuhlsatz, Catherine Riegle-Crumb, and Andrei Cimpian. “Sex and Gender Essentialism in Textbooks.” *Science* 383, no. 6685 (February 23, 2024): 822–825. https://doi.org/10.1126/science.adi1188.

Drea, C. M., N. J. Place, M. L. Weldele, E. M. Coscia, P. Licht, and S. E. Glickman. “Exposure to Naturally Circulating Androgens during Foetal Life Incurs Direct Reproductive Costs in Female Spotted Hyenas, but Is Prerequisite for Male Mating.” *Proceedings of the Royal Society of London. Series B: Biological Sciences* 269, no. 1504 (2002): 1981–1987. https://doi.org/10.1098/rspb.2002.2109.

Duggan, Lisa. *The Twilight of Equality? Neoliberalism, Cultural Politics, and the Attack on Democracy*. Boston: Beacon Press, 2014.

DuVal, Emily H. “Adaptive Advantages of Cooperative Courtship for Subordinate Male Lance-Tailed Manakins.” *American Naturalist* 169, no. 4 (April 2007): 423–432. https://doi.org/10.1086/512137.

DuVal, Emily H., Courtney L. Fitzpatrick, Elizabeth A. Hobson, and Maria R. Servedio. “Inferred Attractiveness: A Generalized Mechanism for Sexual Selection That Can Maintain Variation in Traits and Preferences over Time.” *PLOS Biology* 21, no. 10 (October 2023): e3002269. https://doi.org/10.1371/journal.pbio.3002269.

DuVal, E. H., and J. A. Kapoor. “Causes and Consequences of Variation in Female Mate Search Investment in a Lekking Bird.” *Behavioral Ecology* 26, no. 6 (2015): 1537–1547. https://doi.org/10.1093/beheco/arv110.

DuVal, Emily H., and Bart Kempenaers. “Sexual Selection in a Lekking Bird: The Relative Opportunity for Selection by Female Choice and Male Competition.” *Proceedings of the Royal Society B: Biological Sciences* 275, no. 1646 (2008): 1995–2003. https://doi.org/10.1098/rspb.2008.0151.

Ebaugh, Helen Rose, and Mary Curry. "Fictive Kin as Social Capital in New Immigrant Communities." *Sociological Perspectives* 43, no. 2 (2000): 189–209. https://doi.org/10.2307/1389793.

Emlen, Douglas J. *Animal Weapons: The Evolution of Battle*. New York: Henry Holt, 2014.

The Eugenics Archives. "Intelligence and IQ Testing." Social Sciences and Humanities Research Council of Canada, 2015. https://www.eugenicsarchive.ca/encyclopedia?id=535eecb77095aa000000023a.

Fanon, Frantz. "Colonial Violence and Mental Disorders." In *The Wretched of the Earth*, translated by Richard Philcox. New York: Grove Press, 1961.

Fitzgerald, Terence D. *The Tent Caterpillars*. Ithaca, NY: Cornell University Press, 1995.

Forbes, Jack D. *Columbus and Other Cannibals: The Wétiko Disease of Exploitation, Imperialism, and Terrorism*. Rev. ed. New York: Seven Stories Press, 1978.

Frank, Laurence G., Mary L. Weldele, and Stephen E. Glickman. "Masculinization Costs in Hyaenas." *Nature* 377, no. 6550 (1995): 584–585. https://doi.org/10.1038/377584b0.

Friedner, Michele, and Karen Weingarten. "Disability as Diversity: A New Biopolitics." *Somatosphere: Science, Medicine and Anthropology*, May 23, 2016. http://somatosphere.net/2016/05/disability-as-diversity-a-new-biopolitics.html.

Fuentes, Agustín. "Here's Why Human Sex Is Not Binary." *Scientific American*, May 1, 2023. https://www.scientificamerican.com/article/heres-why-human-sex-is-not-binary/.

Gadgil, M., and K. C. Malhotra. "Adaptive Significance of the Indian Caste System: An Ecological Perspective." *Annals of Human Biology* 10, no. 5 (1983): 465–477. https://doi.org/10.1080/03014468300006671.

Gilmore, Ruth Wilson. *Golden Gulag: Prisons, Surplus, Crisis, and Opposition in Globalizing California*. Berkeley: University of California Press, 2007.

Gorman, James. "Why Are Dogs So Friendly? The Answer May Be in 2 Genes." *New York Times*, July 19, 2017, sec. Science. https://www.nytimes.com/2017/07/19/science/dogs-genes-sociability.html.

Gould, Stephen Jay. *The Mismeasure of Man*. New York: W. W. Norton, 1996.

Gould, Stephen Jay, and Richard C. Lewontin. "The Spandrels of San Marco and the Panglossian Paradigm: A Critique of the Adaptationist Programme." *Proceedings of the Royal Society of London. Series B: Biological Sciences* 205, no. 1161 (1979): 581–598. https://doi.org/10.1098/rspb.1979.0086.

Gowaty, Patricia Adair, Yong-Kyu Kim, and Wyatt W. Anderson. "No Evidence of Sexual Selection in a Repetition of Bateman's Classic Study of *Drosophila melanogaster*." *Proceedings of the National Academy of Sciences* 109, no. 29 (2012): 11740–11745. https://doi.org/10.1073/pnas.1207851109.

Gowaty, Patricia Adair. "Sexual Natures: How Feminism Changed Evolutionary Biology." *Signs: Journal of Women in Culture and Society* 28, no. 3 (2003): 901–921. https://doi.org/10.1086/345324.

Graeber, David, and David Wengrow. *The Dawn of Everything: A New History of Humanity*. New York: Farrar, Straus and Giroux, 2021.

Gruen, Lori. *Entangled Empathy: An Alternative Ethic for Our Relationships with Animals*. New York: Lantern Books, 2015.

Grüter, Christoph, and Walter M. Farina. "The Honeybee Waggle Dance: Can We Follow the Steps?" *Trends in Ecology & Evolution* 24, no. 5 (2009): 242–247. https://doi.org/10.1016/j.tree.2008.12.007.

Gumbs, Alexis Pauline. *Undrowned: Black Feminist Lessons from Marine Mammals*. Chico: AK Press, 2020.

Hajer, Maarten A. *The Politics of Environmental Discourse: Ecological Modernization and the Policy Process*. Oxford: Oxford University Press, 1995.

Hall, Stuart. "Gramsci's Relevance for the Study of Race and Ethnicity." *Journal of Communication Inquiry* (1986): 5–27.

Han, Chang S., and Robert C. Brooks. "Same-Sex Sexual Behaviour as a By-product of Reproductive Strategy under Male–Male Scramble Competition." *Animal Behaviour* 108 (2015): 193–197. https://doi.org/10.1016/j.anbehav.2015.07.035.

Haraway, Donna. "Situated Knowledges: The Science Question in Feminism and the Privilege of Partial Perspective." *Feminist Studies* 14, no. 3 (1988): 575–599. https://doi.org/10.2307/3178066.

Haraway, Donna. *Primate Visions: Gender, Race, and Nature in the World of Modern Science*. New York: Routledge, 1989.

Haraway, Donna J. *When Species Meet*. Minneapolis: University of Minnesota Press, 2008.

Harding, Sandra. "'Strong Objectivity': A Response to the New Objectivity Question." *Synthese* 104, no. 3 (September 1, 1995): 331–349. https://doi.org/10.1007/BF01064504.

Harding, Sandra. *Whose Science? Whose Knowledge? Thinking from Women's Lives*. Ithaca, NY: Cornell University Press, 1991.

Harrington, Anne. *Reenchanted Science: Holism in German Culture from Wilhelm II to Hitler*. Princeton, NJ: Princeton University Press, 1996.

Harris, Macolm. *Palo Alto: A History of California, Capitalism, and the World*. New York: Little, Brown, 2023.

Hayes, Tyrone B., Atif Collins, Melissa Lee, Magdelena Mendoza, Nigel Noriega, A. Ali Stuart, and Aaron Vonk. "Hermaphroditic, Demasculinized Frogs after Exposure to the Herbicide Atrazine at Low Ecologically Relevant Doses." *Proceedings of the National Academy of Sciences* 99, no. 8 (2002): 5476–5480. https://doi.org/10.1073/pnas.082121499.

Hayes, Tyrone B., Vicky Khoury, Anne Narayan, Mariam Nazir, Andrew Park, Travis Brown, Lillian Adame, et al. "Atrazine Induces Complete Feminization and Chemical Castration in Male African Clawed Frogs (*Xenopus laevis*)." *Proceedings of the National Academy of Sciences* 107, no. 10 (2010): 4612–4617. https://doi.org/10.1073/pnas.0909519107.

Heinrich, Bernd. *Ravens in Winter*. New York: Summit Books, 1989.

Henn, Brenna M., Emily Klancher Merchant, Anne O'Connor, and Tina Rulli. "Why DNA Is No Key to Social Equality: On Kathryn Paige Harden's 'The Genetic Lottery.'" *Los Angeles Review of Books*, September 21, 2021, sec. Review. https://lareviewofbooks.org/article/why-dna-is-no-key-to-social-equality-on-kathryn-paige-hardens-the-genetic-lottery/.

Herzig, Rebecca M. *Suffering for Science: Reason and Sacrifice in Modern America*. New Brunswick, NJ: Rutgers University Press, 2006.

Hobbes, Thomas. *Leviathan: Or the Matter, Forme and Power of a Commonwealth, Ecclesiasticall and Civil*. London: Andrew Crooke, 1651.

Horkeimer, Max, and Theodor Adorno. "The Concept of Enlightenment." In *Dialectic of Enlightenment*, translated by John Cumming. New York: Herder & Herder, 1944.

Hrdy, Sarah Blaffer. "Empathy, Polyandry, and the Myth of the Coy Female." In *Feminist Approaches to Science*, edited by Ruth Bleier, 119–146. New York: Pergamon Press, 1986.

Hubbard, Ruth. "The Political Nature of 'Human Nature.'" In *Feminist Frameworks: Alternative Theoretical Accounts of the Relationships Between Men and Women*, edited by Alison Jagger and Paula Rothenberg, 3rd ed. New York: McGraw-HIll, 1993.

Hubbard, Ruth. "Science, Facts, and Feminism." *Hypatia* 3, no. 1 (1988): 5–17.

Hussain, Grace. "The 30 Most Intelligent Animals in the World Might Surprise You." Sentient, September 20, 2023. https://sentientmedia.org/which-animals-are-most-intelligent/.

Imgur. "No, Things Aren't That Simple.—Post." Imgur. Accessed October 19, 2023. https://imgur.com/gallery/1zZ6VSe.

Incorvaia, Darren. "A Two-Ton Lifeguard That Saved a Young Pup." *New York Times*, February 7, 2024, sec. Science. https://www.nytimes.com/2024/02/07/science/elephant-seals-pup-drowning.html.

Irvine, William. *Apes, Angels, and Victorians; the Story of Darwin, Huxley, and Evolution*. New York: McGraw-Hill, 1955.

Janicke, Tim, Ines K. Häderer, Marc J. Lajeunesse, and Nils Anthes. "Darwinian Sex Roles Confirmed across the Animal Kingdom." *Science Advances* 2, no. 2 (2016): e1500983. https://doi.org/10.1126/sciadv.1500983.

Jenkins, Destin, and Justin Leroy, eds. *Histories of Racial Capitalism*. New York: Columbia University Press, 2021.

Jones, Blake Carlton, and Emily H. DuVal. "Mechanisms of Social Influence: A Meta-Analysis of the Effects of Social Information on Female Mate Choice Decisions." *Frontiers in Ecology and Evolution* 7 (2019). https://www.frontiersin.org/articles/10.3389/fevo.2019.00390.

Jordan, David Starr. *The Blood of the Nation: A Study of the Decay of Races Through Survival of the Unfit*. Boston: American Unitarian Association, 1902.

Kafer, Alison. *Feminist, Queer, Crip*. Bloomington: Indiana University Press, 2013.

Kamath, Ambika, and Ashton Wesner. "Animal Territoriality, Property and Access: A Collaborative Exchange between Animal Behaviour and the Social Sciences." *Animal Behaviour* (2020). https://doi.org/10.1016/j.anbehav.2019.12.009.

Kamath, Ambika, and Jonathan Losos. "The Erratic and Contingent Progression of Research on Territoriality: A Case Study." *Behavioral Ecology and Sociobiology* 71, no. 6 (2017): 89. https://doi.org/10.1007/s00265-017-2319-z.

Karkazis, Katrina, Rebecca Jordan-Young, Georgiann Davis, and Silvia Camporesi. "Out of Bounds? A Critique of the New Policies on Hyperandrogenism in Elite Female Athletes." *American Journal of Bioethics* 12, no. 7 (2012): 3–16. https://doi.org/10.1080/15265161.2012.680533.

Kevles, Daniel J. *In the Name of Eugenics: Genetics and the Uses of Human Heredity*. Cambridge, MA: Harvard University Press, 1995.

Kim, Claire Jean. *Dangerous Crossings: Race, Species, and Nature in a Multicultural Age*. Cambridge: Cambridge University Press, 2015. https://doi.org/10.1017/CBO9781107045392.

Kimmerer, Robin Wall. *Braiding Sweetgrass: Indigenous Wisdom, Scientific Knowledge, and the Teachings of Plants*. Minneapolis: Milkweed Editions, 2013.

Kimmerer, Robin Wall. "The Covenant of Reciprocity." In *The Wiley Blackwell Companion to Religion and Ecology*, edited by John Hart, 368–381. Hoboken, NJ: John Wiley & Sons, 2017.

Kimmerer, Robin Wall. "The Serviceberry: An Economy of Abundance." *Emergence Magazine*, October 26, 2022. https://emergencemagazine.org/essay/the-serviceberry/.

King, Erika, Megan V. McPhee, Scott C. Vulstek, Curry J. Cunningham, Joshua R. Russell, and David A. Tallmon. "Alternative Life-History Strategy Contributions to Effective Population Size in a Naturally Spawning Salmon Population." *Evolutionary Applications* 16, no. 8 (2023): 1472–1482. https://doi.org/10.1111/eva.13580.

Ko, Aph. *Racism as Zoological Witchcraft: A Guide for Getting Out*. Brooklyn: Lantern Books, 2019.

Kohn, Eduardo. *How Forests Think: Toward an Anthropology beyond the Human*. Berkeley: University of California Press, 2013.

Kokko, Hanna. "Treat 'em Mean, Keep 'em (Sometimes) Keen: Evolution of Female Preferences for Dominant and Coercive Males." *Evolutionary Ecology* 19, no. 2 (2005): 123–135. https://doi.org/10.1007/s10682-004-7919-1.

Kokko, Hanna, and Johanna Mappes. "Multiple Mating by Females Is a Natural Outcome of a Null Model of Mate Encounters." *Entomologia Experimentalis et Applicata* 146, no. 1 (2013): 26–37. https://doi.org/10.1111/j.1570-7458.2012.01296.x.

Koshyk, Craig. *Pointing Dogs, Volume Two: The British and Irish Breeds*. Winnipeg: Dog Willing, 2023.

Kropotkin, Peter. *Mutual Aid: An Illuminated Factor of Evolution*. Toronto: PM Press, 2021.

Krugman, Paul, and Robin Wells. *Microeconomics*. 2nd ed. New York: Worth Publishers, 2009.

Kuhn, Thomas S. *The Structure of Scientific Revolutions*. 2nd ed. Chicago: University of Chicago Press, 1970.

Kunz, Thomas H., and Wendy R. Hood. "Parental Care and Postnatal Growth in the Chiroptera." In *Reproductive Biology of Bats*, edited by Elizabeth G. Crichton and Philip H. Krutzsch, 415–468. London: Academic Press, 2000.

Lambert, Max R., Geoffrey S. J. Giller, Larry B. Barber, Kevin C. Fitzgerald, and David K. Skelly. "Suburbanization, Estrogen Contamination, and Sex Ratio in Wild Amphibian Populations." *Proceedings of the National Academy of Sciences* 112, no. 38 (2015): 11881–11886. https://doi.org/10.1073/pnas.1501065112.

Lambert, Max R., Tariq Ezaz, and David K. Skelly. "Sex-Biased Mortality and Sex Reversal Shape Wild Frog Sex Ratios." *Frontiers in Ecology and Evolution* 9 (2021): 737. https://doi.org/10.3389/fevo.2021.756476.

Lambert, Max R., Tien Tran, Andrzej Kilian, Tariq Ezaz, and David K. Skelly. "Molecular Evidence for Sex Reversal in Wild Populations of Green Frogs (*Rana clamitans*)." *PeerJ* 7 (2019): e6449. https://doi.org/10.7717/peerj.6449.

Laqueur, Thomas. *Making Sex: Body and Gender from the Greeks to Freud.* Cambridge, MA: Harvard University Press, 1992.

Laskow, Sarah. "The Sad Sex Lives of Suburban Frogs." *Atlas Obscura*, September 9, 2015. http://www.atlasobscura.com/articles/the-sad-sex-lives-of-suburban-frogs.

Laster Pirtle, Whitney N. "Racial Capitalism: A Fundamental Cause of Novel Coronavirus (COVID-19) Pandemic Inequities in the United States." *Health Education & Behavior* 47, no. 4 (August 1, 2020): 504–508. https://doi.org/10.1177/1090198120922942.

Levins, Richard, and Richard C. Lewontin. *The Dialectical Biologist.* Cambridge, MA: Harvard University Press, 1985.

Lewis, Allen R. "Selection of Nuts by Gray Squirrels and Optimal Foraging Theory." *American Midland Naturalist* 107, no. 2 (1982): 250–257. https://doi.org/10.2307/2425376.

Lewis, William. "Louis Althusser." In *The Stanford Encyclopedia of Philosophy*, edited by Edward N. Zalta and Uri Nodelman, Fall 2022. Metaphysics Research Lab, Stanford University, 2022. https://plato.stanford.edu/archives/fall2022/entries/althusser/.

Lewontin, Richard C. "An Estimate of Average Heterozygosity in Man." *American Journal of Human Genetics* 19, no. 5 (1967): 681–685.

Lewontin, Richard C., and Richard Levins. *Biology under the Influence: Dialectical Essays on Ecology, Agriculture, and Health.* New York: Monthly Review Press, 2007.

Leyton, Cristian Alvarado. "Ritual and Fictive Kinship." In *The International Encyclopedia of Anthropology*, edited by Hilary Callan, 1–3. Hoboken, NJ: Wiley, 2018.

Lipshutz, Sara E, and Kimberly A Rosvall. "Neuroendocrinology of Sex-Role Reversal." *Integrative and Comparative Biology* 60, no. 3 (2020): 692–702. https://doi.org/10.1093/icb/icaa046.

Lloyd, Elisabeth A. "Adaptationism and the Logic of Research Questions: How to Think Clearly about Evolutionary Causes." *Biological Theory* 10, no. 4 (2015): 343–362. https://doi.org/10.1007/s13752-015-0214-2.

Longino, Helen. "Does the Structure of Scientific Revolutions Permit a Feminist Revolution in Science?" In *Thomas Kuhn*, edited by Thomas Nickles, 261–281. New York: Cambridge University Press, 2002.

Losos, Jonathan. *Lizards in an Evolutionary Tree: Ecology and Adaptive Radiation of Anoles*. Oakland: University of California Press, 2011.

Lowe, Lisa. *The Intimacies of Four Continents*. Durham, NC: Duke University Press, 2015.

Lugones, María. "Heterosexualism and the Colonial/Modern Gender System." *Hypatia* 22, no. 1 (2007): 186–219. https://doi.org/10.1111/j.1527-2001.2007.tb01156.x.

MacArthur, Robert H., and Eric R. Pianka. "On Optimal Use of a Patchy Environment." *American Naturalist* 100, no. 916 (1966): 603–609. https://doi.org/10.1086/282454.

MacDonald, Ann-Marie, Marc De Guerre, and Alan Mendelsohn. *The Disappearing Male*. Documentary. Toronto: Canadian Broadcasting Corporation, 2008.

Mann, Charles C. "The Book That Incited a Worldwide Fear of Overpopulation." *Smithsonian Magazine*. Accessed February 26, 2024. https://www.smithsonianmag.com/innovation/book-incited-worldwide-fear-overpopulation-180967499/.

Marino, Lori. "Thinking Chickens: A Review of Cognition, Emotion, and Behavior in the Domestic Chicken." *Animal Cognition* 20, no. 2 (2017): 127–147. https://doi.org/10.1007/s10071-016-1064-4.

Markowitz, Sally. "Pelvic Politics: Sexual Dimorphism and Racial Difference." *Signs* 26, no. 2 (2001): 389–414.

Martin, Emily. "The Egg and the Sperm: How Science Has Constructed a Romance Based on Stereotypical Male-Female Roles." *Signs* 16, no. 3 (1991): 485–501.

Mau, Søren. *Mute Compulsion: A Marxist Theory of the Economic Power of Capital*. London: Verso, 2023.

Maung, Hane Htut. "Classifying Sexes." *Journal of Diversity and Gender Studies* 10, no. 1 (2023): 35–52.

Mayr, Ernst. "How to Carry Out the Adaptationist Program?" *American Naturalist* 121, no. 3 (March 1983): 324–334. https://doi.org/10.1086/284064.

McLaughlin, J. F., Kinsey M. Brock, Isabella Gates, Anisha Pethkar, Marcus Piattoni, Alexis Rossi, and Sara E. Lipshutz. "Multivariate Models of Animal Sex: Breaking Binaries Leads to a Better Understanding of Ecology and Evolution." *Integrative and Comparative Biology* 63, no. 4 (2023): 891–906. https://doi.org/10.1093/icb/icad027.

Medicine, Beatrice. "Directions in Gender Research in American Indian Societies: Two Spirits and Other Categories." *Online Readings in Psychology and Culture* 3, no. 1 (August 1, 2002). https://doi.org/10.9707/2307-0919.1024.

Milam, Erika L. *Looking for a Few Good Males: Female Choice in Evolutionary Biology*. Baltimore: Johns Hopkins University Press, 2010.

Miller, Lulu. *Why Fish Don't Exist: A Story of Loss, Love, and the Hidden Order of Life*. New York: Simon & Schuster, 2021.

Mitman, Gregg. *The State of Nature: Ecology, Community, and American Social Thought, 1900–1950*. Chicago: University of Chicago Press, 1992.

Miyagi, Miriam, Eartha Mae Guthman, and Simón(e) Dow-Kuang Sun. "Transgender Rights Rely on Inclusive Language." *Science* 374, no. 6575 (2021): 1568–1569. https://doi.org/10.1126/science.abn3759.

Molteni, Megan. "Buffalo Shooting Ignites a Debate over the Role of Genetics Researchers in White Supremacist Ideology." *STAT* (blog), May 23, 2022. https://www.statnews.com/2022/05/23/buffalo-shooting-ignites-debate-genetics-researchers-in-white-supremacist-ideology/.

Monk, Julia D., Erin Giglio, Ambika Kamath, Max R. Lambert, and Caitlin E. McDonough. "An Alternative Hypothesis for the Evolution of Same-Sex Sexual Behaviour in Animals." *Nature Ecology & Evolution* (2019): 1–10. https://doi.org/10.1038/s41559-019-1019-7.

Moore, David S., and David Shenk. "The Heritability Fallacy." *WIREs Cognitive Science* 8, nos. 1–2 (2017): e1400. https://doi.org/10.1002/wcs.1400.

Mortimer-Sandilands, Catriona, and Bruce Erickson, eds. *Queer Ecologies: Sex, Nature, Politics, Desire*. Bloomington: Indiana University Press, 2010.

Muir, W. M. "Group Selection for Adaptation to Multiple-Hen Cages: Selection Program and Direct Responses." *Poultry Science* 75, no. 4 (1996): 447–458. https://doi.org/10.3382/ps.0750447.

Munger, James C. "Optimal Foraging? Patch Use by Horned Lizards (Iguanidae: *Phrynosoma*)." *American Naturalist* 123, no. 5 (1984): 654–680. https://doi.org/10.1086/284230.

Murphy, Michelle. *The Economization of Life*. Durham, NC: Duke University Press, 2017.

Murphy, Michelle. *Sick Building Syndrome and the Problem of Uncertainty: Environmental Politics, Technoscience, and Women Workers*. Durham, NC: Duke University Press, 2006.

Nagel, Thomas. *The View from Nowhere*. Oxford: Oxford University Press, 1989.

Nelson, Alondra. *The Social Life of DNA*. Boston: Beacon Press, 2016.

Noble, G. K., and H. T. Bradley. "The Mating Behavior of Lizards; Its Bearing on the Theory of Sexual Selection." *Annals of the New York Academy of Sciences* 35, no. 1 (1933): 25–100. https://doi.org/10.1111/j.1749-6632.1933.tb55365.x.

Ogbunugafor, C. Brandon. "DNA, Basketball, and Birthday Luck. A Review of *The Genetic Lottery: Why DNA Matters for Social Equality*." *American Journal of Biological Anthropology* 179, no. 3 (2022): 501–504. https://doi.org/10.1002/ajpa.24599.

Okasha, Samir. *Agents and Goals in Evolution*. Oxford: Oxford University Press, 2018.

Oudshoorn, Nelly. *Beyond the Natural Body: An Archaeology of Sex Hormones*. New York: Routledge, 1994.

Oyěwùmí, Oyèrónkẹ́. *The Invention of Women: Making an African Sense of Western Gender Discourses*. Minneapolis: University of Minnesota Press, 1997.

Packer, Melina, and Max R. Lambert. "What's Gender Got to Do with It? Dismantling the Human Hierarchies in Evolutionary Biology and Environmental Toxicology for Scientific and Social Progress." *American Naturalist* 200, no. 1 (2022): 114–128. https://doi.org/10.1086/720131.

Panofsky, Aaron. *Misbehaving Science: Controversy and the Development of Behavior Genetics*. Chicago: University of Chicago Press, 2014.

Pettersson, Irina, and Cecilia Berg. "Environmentally Relevant Concentrations of Ethynylestradiol Cause Female-Biased Sex Ratios in *Xenopus tropicalis* and *Rana temporaria*." *Environmental Toxicology and Chemistry* 26, no. 5 (2007): 1005–1009. https://doi.org/10.1897/06-464r.1.

Piepzna-Samarasinha, Leah Lakshmi. *Care Work: Dreaming Disability Justice*. Vancouver: Arsenal Pulp Press, 2018.

Pierce, G. J., and J. G. Ollason. "Eight Reasons Why Optimal Foraging Theory Is a Complete Waste of Time." *Oikos* 49, no. 1 (1987): 111–118. https://doi.org/10.2307/3565560.

Pollock, Anne. "Queering Endocrine Disruption." In *Object-Oriented Feminism*, edited by Katherine Behar, 183–199. Minneapolis: University of Minnesota Press, 2016.

Prescod-Weinstein, Chanda, and Katherine McKittrick. "Public Thinker: Katherine McKittrick on Black Methodologies and Other Ways of Being." *Public Books* (blog), February 1, 2021. https://www.publicbooks.org/public-thinker-katherine-mckittrick-on-black-methodologies-and-other-ways-of-being/.

Reardon, Jenny. *Race to the Finish: Identity and Governance in an Age of Genomics*. Princeton, NJ: Princeton University Press, 2005.

Richardson, Sarah S. "Sex Contextualism." *Philosophy, Theory, and Practice in Biology* 14, no. 2 (2022). https://doi.org/10.3998/ptpbio.2096.

Richardson, Sarah S. "Sexing the X: How the X Became the 'Female Chromosome.'" *Signs* 37, no. 4 (2012): 909–933. https://doi.org/10.1086/664477.

Riskin, Jessica. *The Restless Clock: A History of the Centuries-Long Argument over What Makes Living Things Tick*. Chicago: University of Chicago Press, 2018.

Ritvo, Harriet. *The Animal Estate: The English and Other Creatures in the Victorian Age*. Cambridge, MA: Harvard University Press, 1987.

Roberts, Dorothy E. "Is Race-Based Medicine Good for Us? African American Approaches to Race, Biomedicine, and Equality." *Journal of Law, Medicine & Ethics* 36, no. 3 (2008): 537–545. https://doi.org/10.1111/j.1748-720X.2008.302.x.

Robinson, Cedric J. *Black Marxism: The Making of the Black Radical Tradition*. Chapel Hill: University of North Carolina Press, 2000.

Roughgarden, Joan. *Evolution's Rainbow: Diversity, Gender, and Sexuality in Nature and People*. 10th ann. ed. Berkeley: University of California Press, 2013.

Roy, Deboleena. *Molecular Feminisms: Biology, Becomings, and Life in the Lab*. Seattle: University of Washington Press, 2018.

Rubenstein, Dustin R., and John Alcock. *Animal Behavior*. 11th ed. Sunderland, MA: Sinauer Associates, 2019.

Russett, Cynthia E. *Sexual Science: The Victorian Construction of Womanhood*. Cambridge, MA: Harvard University Press, 1995.

Ryang, Sonia. "A Note on Transnational Consanguinity, or, Kinship in the Age of Terrorism." *Anthropological Quarterly* 77, no. 4 (2004): 747–770.

Sahlins, Marshall. *The Use and Abuse of Biology: An Anthropological Critique of Sociobiology*. Ann Arbor: University of Michigan Press, 1977.

Saini, Angela. *Inferior: How Science Got Women Wrong—and the New Research That's Rewriting the Story*. Boston: Beacon Press, 2017.

Salam, Erum. "Canada Issues Travel Advisory for LGBTQ+ Residents Visiting US." *The Guardian*, August 31, 2023, sec. World News. https://www.theguardian.com/world/2023/aug/31/canada-travel-warning-lgbtq-residents-us.

Salmón, Enrique. "Kincentric Ecology: Indigenous Perceptions of the Human-Nature Relationship." *Ecological Applications* 10, no. 5 (2000): 1327–1332.

Sapolsky, Robert M. *The Trouble with Testosterone: And Other Essays on the Biology of the Human Predicament*. New York: Touchstone, 1998.

Sardell, Rebecca J., Bart Kempenaers, and Emily H. DuVal. "Female Mating Preferences and Offspring Survival: Testing Hypotheses on the Genetic Basis of Mate Choice in a Wild Lekking Bird." *Molecular Ecology* 23, no. 4 (2014): 933–946. https://doi.org/10.1111/mec.12652.

Scharf, Inon, and Oliver Y. Martin. "Same-Sex Sexual Behavior in Insects and Arachnids: Prevalence, Causes, and Consequences." *Behavioral Ecology and Sociobiology* 67, no. 11 (2013): 1719–1730. https://doi.org/10.1007/s00265-013-1610-x.

Schrefer, Eliot. *Queer Ducks (and Other Animals): The Natural World of Animal Sexuality*. New York: Katherine Tegen Books, 2022.

Schuller, Kyla. *The Biopolitics of Feeling: Race, Sex, and Science in the Nineteenth Century*. Durham, NC: Duke University Press, 2017.

Seth, Suman. "Darwin and the Ethnologists: Liberal Racialism and the Geological Analogy." *Historical Studies in the Natural Sciences* 46, no. 4 (2016): 490–527. https://doi.org/10.1525/hsns.2016.46.4.490.

Sewell, William H., Jr. *Logics of History: Social Theory and Social Transformation*. Chicago: University of Chicago Press, 2005.

Seymour, Nicole. *Strange Natures: Futurity, Empathy, and the Queer Ecological Imagination*. Urbana: University of Illinois Press, 2013.

Shavit, Ayelet. "Shifting Values Partly Explain the Debate over Group Selection." *Studies in History and Philosophy of Science Part C: Studies in History and Philosophy of Biological and Biomedical Sciences* 35, no. 4 (2004): 697–720. https://doi.org/10.1016/j.shpsc.2004.09.007.

Shotwell, Alexis. *Against Purity: Living Ethically in Compromised Times*. Minneapolis: University of Minnesota Press, 2016.

Simmons, Leigh W., and T. James Ridsdill-Smith. *Ecology and Evolution of Dung Beetles*. Oxford: John Wiley & Sons, 2011.

Skidelsky, Robert. *What's Wrong with Economics? A Primer for the Perplexed*. New Haven, CT: Yale University Press, 2020.

Snyder, Brian F., and Patricia Adair Gowaty. "A Reappraisal of Bateman's Classic Study of Intrasexual Selection." *Evolution* 61, no. 11 (2007): 2457–2468.

Spade, Dean. "Solidarity Not Charity: Mutual Aid for Mobilization and Survival." *Social Text* 38, no. 1 (142) (March 1, 2020): 131–151. https://doi.org/10.1215/01642472-7971139.

Stack, Carol B. *All Our Kin: Strategies for Survival in a Black Community*. 1974. Reprint, New York: Basic Books, 2003.

Starr, Kelsey Jo, and Neha Sahgal. "Measuring Caste in India." *Decoded* (blog), June 29, 2021. https://www.pewresearch.org/decoded/2021/06/29/measuring-caste-in-india/.

Stern, Alexandra Minna. *Eugenic Nation: Faults and Frontiers of Better Breeding in Modern America*. 2nd ed. Berkeley: University of California Press, 2015.

Strathern, Marilyn. *Kinship, Law and the Unexpected: Relatives Are Always a Surprise*. New York: Cambridge University Press, 2005.

Stuart, Yoel E., Thor Veen, Jesse N. Weber, Dieta Hanson, Mark Ravinet, Brian K. Lohman, Cole J. Thompson, et al. "Contrasting Effects of Environment and Genetics Generate a Continuum of Parallel Evolution." *Nature Ecology & Evolution* 1, no. 6 (2017): 1–7. https://doi.org/10.1038/s41559-017-0158.

Subramaniam, Banu. "Snow Brown and the Seven Detergents: A Metanarrative on Science and the Scientific Method." *Women's Studies Quarterly* 28, no. 1/2 (2000): 296–304.

Sultan, Sonia E., Armin P. Moczek, and Denis Walsh. "Bridging the Explanatory Gaps: What Can We Learn from a Biological Agency Perspective?" *BioEssays* 44, no. 1 (2022): 2100185. https://doi.org/10.1002/bies.202100185.

Svensson, Erik I. "On Reciprocal Causation in the Evolutionary Process." *Evolutionary Biology* 45, no. 1 (2018): 1–14. https://doi.org/10.1007/s11692-017-9431-x.

TallBear, Kimberly. *Native American DNA: Tribal Belonging and the False Promise of Genetic Science*. Minneapolis: University of Minnesota Press, 2013.

TallBear, Kimberly. "Why Interspecies Thinking Needs Indigenous Standpoints." *Cultural Anthropology*, April 24, 2011. https://culanth.org/fieldsights/260-why-interspecies-thinking-needs-indigenous-standpoints.

Tang-Martínez, Zuleyma. "Rethinking Bateman's Principles: Challenging Persistent Myths of Sexually Reluctant Females and Promiscuous Males." *Journal of Sex Research* 53, no. 4–5 (May 3, 2016): 532–559. https://doi.org/10.1080/00224499.2016.1150938.

Tang-Martinez, Zuleyma, and T. Brandt Ryder. "The Problem with Paradigms: Bateman's Worldview as a Case Study." *Integrative and Comparative Biology* 45, no. 5 (2005): 821–830. https://doi.org/10.1093/icb/45.5.821.

Taylor, Dorceta E. *The Rise of the American Conservation Movement: Power, Privilege, and Environmental Protection*. Durham, NC: Duke University Press, 2016.

Taylor, Sunaura. *Beasts of Burden: Animal and Disability Liberation*. New York: New Press, 2017.

Thomas, Renny. "Brahmins as Scientists and Science as Brahmins' Calling: Caste in an Indian Scientific Research Institute." *Public Understanding of Science* (2020): 306–318. https://doi.org/10.1177/0963662520903690.

Thompson, Fred G. "Notes on the Behavior of the Lizard *Anolis carolinensis*." *Copeia* 1954, no. 4 (1954): 299. https://doi.org/10.2307/1440053.

Tiffany P. "Crow Uses Plastic Lid to Sled Down Roof Over and Over Again." YouTube, August 31, 2014. https://www.youtube.com/watch?v=L9mrTdYhOHg.

Trivers, Robert L. "Parental Investment and Sexual Selection." In *Sexual Selection and the Descent of Man*, edited by Bernard Campbell, 35–57. Chicago: Aldine, 1972.

Velocci, Beans. "Binary Logic." Yale University, 2021.

Waddington, Keith D., and Larry R. Holden. "Optimal Foraging: On Flower Selection by Bees." *American Naturalist* 114, no. 2 (1979): 179–196. https://doi.org/10.1086/283467.

Wade, Michael J. *Adaptation in Metapopulations: How Interaction Changes Evolution*. Chicago: University of Chicago Press, 2016.

Wadsack, Justine, Rachel Jones, and Cory McGarr. "Gender Transition; Prohibitions; Hormone Therapies," Pub. L. No. SB1702 (2023). https://www.azleg.gov/legtext/56leg/1R/bills/SB1702P.pdf.

Watters, Jason V. "Can the Alternative Male Tactics 'Fighter' and 'Sneaker' Be Considered 'Coercer' and 'Cooperator' in Coho Salmon?" *Animal Behaviour* 70, no. 5 (2005): 1055–1062. https://doi.org/10.1016/j.anbehav.2005.01.025.

"Wayfinding." Accessed October 20, 2023. https://archive.hokulea.com/navigate/navigate.html.

Weaver, Harlan. "Pit Bull Promises: Inhuman Intimacies and Queer Kinships in an Animal Shelter." *GLQ: A Journal of Lesbian and Gay Studies* 21, no. 2 (2015): 343–363.

Weston, Kath. *Families We Choose: Lesbians, Gays, Kinship*. New York: Columbia University Press, 1997.

"When the Strong Outbreed the Weak: An Interview with William Muir—This View of Life." July 11, 2016. https://thisviewoflife.com/when-the-strong-outbreed-the-weak-an-interview-with-william-muir/.

Williams, George C. *Adaptation and Natural Selection: A Critique of Some Current Evolutionary Thought*. Princeton Science Library ed. Princeton, NJ: Princeton University Press, 1996.

Williams, George C. "A Sociobiological Expansion of Evolution and Ethics." In *Evolution and Ethics: T.H. Huxley's Evolution and Ethics with New Essays on Its Victorian and Sociobiological Context*, edited by George Christopher Williams and James G. Paradis, 179–214. Princeton, NJ: Princeton University Press, 1989.

Wilson, David Sloan. *Does Altruism Exist? Culture, Genes, and the Welfare of Others*. New Haven, CT: Yale University Press, 2015.

Wilson, David Sloan. "Altruism and Organism: Disentangling the Themes of Multilevel Selection Theory." *American Naturalist* 150, no. S1 (July 1997): S122–S34. https://doi.org/10.1086/286053.

Wilson, Edward O. *Sociobiology: The New Synthesis*. Cambridge, MA: Belknap Press of Harvard University Press, 1975.

Wynne-Edwards, V. C. *Animal Dispersion in Relation to Social Behaviour*. Edinburgh: Oliver and Boyd, 1962.

Wynne-Edwards, V. C. "Population Control in Animals." *Scientific American* 211, no. 2 (1964): 68–75.

Wynne-Edwards, V. C. "Self-Regulating Systems in Populations of Animals: A New Hypothesis Illuminates Aspects of Animal Behavior That Have Hitherto Seemed Unexplainable." *Science* 147, no. 3665 (1965): 1543–1548. https://doi.org/10.1126/science.147.3665.1543.

Yong, Ed. "The Genes That Built a Home." *National Geographic*, January 16, 2013. https://www.nationalgeographic.com/science/article/genetics-burrowing.

Yong, Ed. *An Immense World: How Animal Senses Reveal the Hidden Realms around Us*. New York: Random House, 2022.

Index

Page numbers followed by "f" indicate figures.